THE ELEMENTS

OF

PLANE ANALYTIC GEOMETRY.

A TEXT-BOOK

INCLUDING NUMEROUS EXAMPLES AND APPLICATIONS,
AND ESPECIALLY DESIGNED FOR BEGINNERS.

BY

GEORGE R. BRIGGS,
FORMERLY TUTOR IN MATHEMATICS IN HARVARD UNIVERSITY.

SEVENTH EDITION, REVISED AND ENLARGED

BY

MAXIME BÔCHER,
ASSISTANT PROFESSOR IN HARVARD UNIVERSITY.

NEW YORK:
JOHN WILEY & SONS.
LONDON: CHAPMAN & HALL, LIMITED.
1903.

ROBERT DRUMMOND, PRINTER, NEW YORK.

PREFACE.

THE design of this book is to present the elements of Analytic Geometry in a simple form, for the use of beginners. With this object in view, I have tried to make the demonstrations thorough, sometimes at the expense of brevity, in order that the student may never feel that the fundamental truths have been in part assumed. Therefore it may often seem best to the teacher to present special cases of the theorems as preparation for the general demonstrations of the book.

The text has been interspersed with a large number of examples and applications to familiar theorems of Geometry, often accompanied with the answers and hints for the solution, to stimulate the interest of the student and to teach him early how to work for himself. For the same purpose the chapter on Loci has been introduced.

Although this book has been prepared for the use of the Freshman Class in Harvard College, it is hoped that it may prove useful in Schools and Academies, and to any student having a fair knowledge of Algebra and Trigonometry who wishes to prepare for a more extended course in Analytic Geometry.

I gladly acknowledge my indebtedness to Prof. Byerly for his kindness in reading the manuscript and for valuable suggestions. I have also freely used the works of Salmon, Todhunter, Puckle, and others.

G. R. BRIGGS.

CAMBRIDGE, April, 1881.

THE changes in the present edition consist mainly in the addition of a moderate amount of new matter. Almost all of this will be found after page 107, and in particular the last half of Chapter VIII and the whole of Chapter IX are new. It is hoped that in its present form the book will be found to cover the ground of the ordinary first course in analytic geometry as given in American colleges and technical schools, and that it will prove serviceable as a text-book in such courses.

MAXIME BÔCHER.

CAMBRIDGE, January, 1903.

CONTENTS.

ELEMENTS OF ANALYTIC GEOMETRY.

CHAPTER I.

INTRODUCTION. — THE POINT.

1. THE reader is probably familiar with algebraic statements of geometrical theorems which concern the magnitude of lines; for example, where a, b, and c represent the legs and hypotenuse of a right triangle, $a^2 + b^2 = c^2$ is a statement in algebraic language of the Pythagorean theorem.

There is a large class of geometrical theorems which concern the position of lines; for example, a tangent to a circle is perpendicular to the radius drawn to the point of tangency. Analytic Geometry enables us to apply algebra to this class of problems; and as we advance with our subject we shall see that this use of algebra shortens our work and generalizes our reasoning.

2. We shall assume that the reader is acquainted with the application of the algebraic signs *plus* and *minus* to express the *upward* and *downward* directions of vertical lines, the directions *from left to right* and *from right to left* of horizontal lines, and, in general, to distinguish between the *opposite* directions of lines which have the same general direction. The following theorem, a direct result of the distinction between positive and negative lines, will be useful for reference: —

If three points, A, B, and C, lie on the same straight line, we may write, when we consider the directions as well as the lengths of lines,

$$AB + BC = AC,$$

whatever the arrangement of these points on that line.

3. Analogous to the theorem of § 2, there is a principle for the addition of angles which depends upon the distinction between opposite rotations by the signs *plus* and *minus*. We shall place it here for reference.

If three lines, OA, OB, and OC, pass through the same point O, we may write

$$AOB + BOC = AOC,$$

whatever the arrangement of the three lines which pass through O.

4. In considering problems which concern the position of points and lines in a plane, we must have some fixed objects in the plane to which we may refer the position of each point and line. There are several combinations of fixed objects which are used for this purpose, the most common being two straight lines at right angles to each other and indefinite in extent.

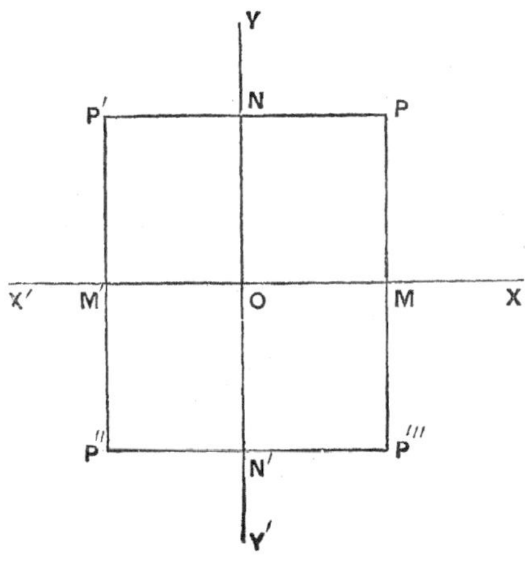

In the figure, $X'X$ and $Y'Y$ are the fixed straight lines intersecting at O. The position of P is determined if we know NP, its distance *from* $Y'Y$, and MP, its distance *from* $X'X$, *together with the directions of these lines;* for we know from geometry that P is the only point in the plane at a distance NP *to the right of* $Y'Y$ and at a distance MP

above $X'X$. Let P, P', P'', and P''' be the corners of a rectangle whose middle point is O, and whose sides are parallel to $X'X$ and $Y'Y$. Then recollecting the conventions of § 2, if we represent the length NP by a and the length MP by b, the distances of P from $Y'Y$ and $X'X$ respectively are a and b; of P', $-a$ and b; of P'', $-a$ and $-b$; of P''', a and $-b$.

5. We see that two elements determine the position of each point. These are the straight lines which indicate by their lengths the distances of the point from $X'X$ and $Y'Y$, and by their signs its directions *from* these lines. These elements are called the *coördinates* of the point which they determine: the one which indicates the distance and direction from $Y'Y$ is called the *abscissa*; the other, which gives the distance and direction from $X'X$, is called the *ordinate.*

In the figure, NP, or its equal both in length and direction, OM, is the abscissa of P, MP is the ordinate of P; or we may say that the abscissa of P is a, and the ordinate b; or, more briefly, that the coördinates of P are a and b, the abscissa always being named first.

The fixed lines are called *axes* of coördinates, and we speak of $X'X$ as the *axis* of x, or of abscissas, while $Y'Y$ is called the *axis* of y, or of ordinates. O is called the *origin* of coördinates, or simply the *origin.*

In order to abbreviate the terms *abscissa* and *ordinate*, it is customary to denote them by the letters x and y respectively; i. e. for the point P, $x = a$, $y = b$; for P', $x = -a$, $y = b$, etc.

6. If we call the portions of the plane lying in the angles XOY, YOX', $X'OY'$, $Y'OX$, the *first*, *second*, *third*, and *fourth quadrants* respectively, we readily see that any point in the first quadrant has x positive and y positive; any point in the second quadrant has x negative and y positive; any point in the third quadrant has x negative and y negative; any point

in the fourth quadrant has x positive and y negative. The signs of the coördinates of a point, therefore, show in which quadrant the point lies, while the magnitudes of the coördinates fix the position of the point in that quadrant.

7. In the development of our subject we shall have to consider points which move in the plane and others which are stationary. The coördinates of a moving point are *variables*, for each assumes in turn an indefinite number of values. We shall designate a moving point by P, its coördinates by x and y, and the foot of its ordinate by M.

The fixed points may be either in known positions or unknown. We shall represent known points by P_1, P_2, P_3, etc., their coördinates by (x_1, y_1), (x_2, y_2), (x_3, y_3), etc., and the feet of their ordinates by M_1, M_2, M_3, etc., respectively.

The points whose positions are unknown, yet determined by the conditions of the problem, we shall call P', P'', etc., their coördinates (x', y'), (x'', y''), etc., and the feet of their ordinates M', M'', etc., respectively.

The student should be careful to accustom himself to this notation, as it is very important not to confuse the moving points with those which have determined positions.

Instead of designating a point P_1 as having coördinates $x_1 = a$, $y_1 = b$, we often write the point (a, b), always understanding that the abscissa is written first; e. g. the points $(1, 2)$, $(3, -2)$, are the points P_1, with $x_1 = 1$ and $y_1 = 2$, and P_2, with $x_2 = 3$ and $y_2 = -2$.

8. The system of coördinates described in §§ 4–6 is called *rectangular*, to distinguish it from another system often employed, in which the axes are not perpendicular to each other, but may make any angle. In the latter system the abscissa of a point is measured parallel to the axis of x, and the ordinate, parallel to the axis of y. This system is called *oblique;* and, together with the rectangular system, forms the general class of *rectilinear* systems in distinction from the *polar* system, which will be mentioned later.

We shall always suppose coördinates rectangular unless otherwise stated.

9. It is easy to represent the position of any given point, in a figure drawn to any scale. This is called *plotting* or *constructing* the point.

Let us plot the points (2, 3), (– 2, 4), (– 3, – 2).

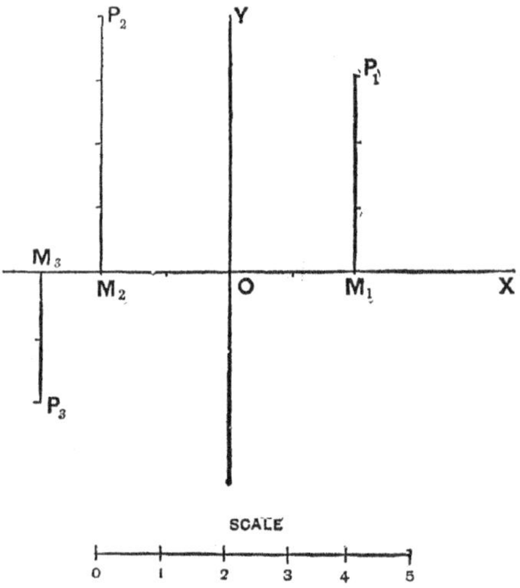

The first point, which we will call P_1, is at a distance of *two* units to the *right* of OY, and *three* units *above* OX, as indicated by its coördinates. We may measure off the first distance *from* O along OX to M_1, which must be the foot of the ordinate of P_1. Now the ordinate must be drawn upward parallel to OY, and made equal to three units of length. Its upper extremity is P_1, the required position.

To plot P_2, we lay off $OM_2 = -2$ and $M_2P_2 = 4$. To plot P_3, we make $OM_3 = -3$ and $M_3P_3 = -2$. In a similar way any point may be plotted or constructed.

EXAMPLES.

(1.) Plot the points (4, – 3), (6, 0), (– 3, 0), (0, 4), (0, – 2), (0, 0).

(2.) Draw the straight line whose extremities are (– 2, 3) and (2, – 3).

(3.) Draw the triangle, the vertices of which are the points $(-4, -3)$, $(6, 1)$, $(4, 11)$.

10. We will now seek an expression for the distance between two given points, in terms of the coördinates of these points.

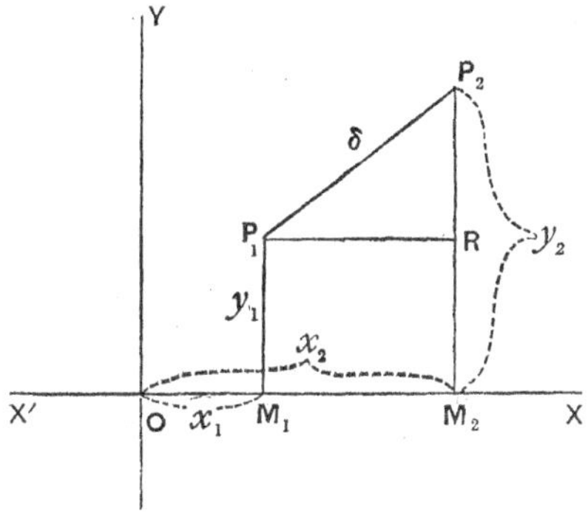

Let P_1, with coördinates $x_1 = O\,M_1$, $y_1 = M_1\,P_1$, and P_2, with coördinates $x_2 = O\,M_2$, $y_2 = M_2\,P_2$, be the given points. Let δ represent the distance between P_1 and P_2. Through P_1 draw a parallel to $O\,X$, and let it meet $M_2\,P_2$ (or this line extended if necessary) in the point R. Now we know that $P_1\,R\,P_2$ must be a right triangle, and therefore that

$$\delta^2 = \overline{P_1\,R}^2 + \overline{R\,P_2}^2 \quad . \; . \; . \; . \; . \; . \; . \; . \; . \; . \; . \quad (1)$$

But $P_1\,R$ is equal to $M_1\,M_2$, the opposite side of a rectangle, both in length and direction. The points M_1, O, and M_2 are by construction on the same straight line $X'\,X$, and therefore, by § 2, we may write

$$\begin{aligned} P_1\,R = M_1\,M_2 = M_1\,O + O\,M_2 &= O\,M_2 - O\,M_1 \\ &= \quad x_2 \quad - \quad x_1 \quad . \; . \; . \; . \quad (2) \end{aligned}$$

Again, R, M_2, and P_2 are by construction on one straight line, and therefore, by § 2,

$$\begin{aligned} R\,P_2 = R\,M_2 + M_2\,P_2 &= M_2\,P_2 - M_2\,R \\ &= M_2\,P_2 - M_1\,P_1 \\ &= y_2 - y_1, \quad . \; . \; . \; . \; . \; . \; . \; . \; . \quad (3) \end{aligned}$$

because $M_2\,R$ and $M_1\,P_1$ are equal, being opposite sides of a rectangle.

Substituting in equation (1) the values of $P_1\,R$ and $R\,P_2$ found in equations (2) and (3), we have

$$\delta^2 = (x_2 - x_1)^2 + (y_2 - y_1)^2,$$

or
$$\delta = \sqrt{(x_2 - x_1)^2 + (y_2 - y_1)^2};$$

an equation which answers the requirements of our problem.

We may interchange P_1 and P_2 in the value of δ just found, as by so doing, though we change the signs of the quantities in the parentheses, we *do not change their squares*.

NOTE.—The student should study the solution of this problem with great care, and satisfy himself that every step in the reasoning is applicable to all the modified forms of the figure which result from taking P_1 and P_2 in different quadrants. If he considers the theorem stated in § 2 until its full meaning becomes clear, he will be convinced of the generality of the solution of this problem. It is important that he be so convinced, as we have here illustrated in a simple form a method of reasoning which will be often employed hereafter.

EXAMPLES.*

(1.) What is the length of the straight line which joins $(7, -2)$ with $(4, -3)$?

Here we may write

$$x_1 = 7, \quad y_1 = -2, \quad x_2 = 4, \quad y_2 = -3.$$

Substituting these values in the expression for δ, we have

$$\delta = \sqrt{[4-7]^2 + [-3-(-2)]^2} = \sqrt{(-3)^2 + (-1)^2}$$
$$= \sqrt{10} \qquad \textit{Ans.}$$

(2.) Find the distance between $(2, 3)$ and $(-2, 8)$.

Ans. $\sqrt{41}$.

(3.) Find the length of the line given in § 9, Ex. (2).

Ans. $2\sqrt{13}$.

(4.) Show that the lengths of the sides of the triangle mentioned in § 9, Ex. (3), are in the same ratio as $\sqrt{26}$, $\sqrt{29}$, $\sqrt{65}$.

(5.) Show that the distance of a given point, P_1, from the origin is $\sqrt{x_1^2 + y_1^2}$.

* In solving examples, the student should construct figures, plotting each point.

(6.) Calculate the lengths of the sides of the rectangle whose vertices are the points (a, b), $(-a, b)$, $(-a, -b)$, $(a, -b)$; show that the vertices are equidistant from the origin; prove that the diagonals are equal.

11. Let us find the coördinates of a point P' so situated on the straight line which passes through two given points, P_1 and P_2, that its distances *from* these points respectively are in a given ratio, $k = \frac{m_1}{m_2}$.

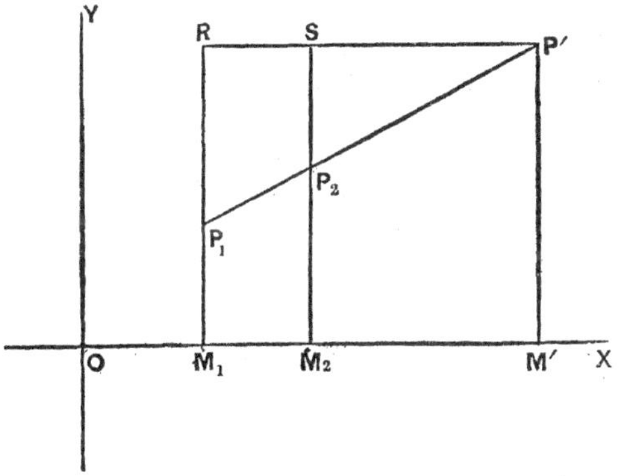

By the conditions of the problem $\frac{P_1P'}{P_2P'} = \frac{m_1}{m_2}$, where m_1 and m_2 are *any two quantities* and have no connection with the *points* M_1 and M_2. Through P' draw a line parallel to OX and cutting M_1P_1 and M_2P_2 (or their extensions) in R and S respectively. Since M_1P_1 and M_2P_2 are necessarily parallel, the triangles P_1RP' and P_2SP' are always similar, and therefore

$$\frac{RP'}{SP'} = \frac{P_1P'}{P_2P'} = \frac{m_1}{m_2} \quad \ldots\ldots\ldots\ldots \quad (1)$$

Now, by § 2,

$$RP' = M_1M' = M_1O + OM' = OM' - OM_1$$
$$= x' - x_1,$$

and similarly

$$SP' = M_2M' = x' - x_2.$$

Substituting these values of RP' and SP' in equation (1), we have

$$\frac{x' - x_1}{x' - x_2} = \frac{m_1}{m_2}.$$

Solving for x',

$$x' = \frac{m_2 x_1 - m_1 x_2}{m_2 - m_1},$$

which gives the value of the unknown abscissa of P' in terms of known quantities.

Similarly we may find y'; for

$$\frac{P_1 R}{P_2 S} = \frac{P_1 P'}{P_2 P'} = \frac{m_1}{m_2} \quad \ldots\ldots\ldots\ldots \quad (2)$$

Now, by § 2,

$$\begin{aligned} P_1 R = P_1 M_1 + M_1 R &= M_1 R - M_1 P_1 \\ &= M' P' - M_1 P_1 \\ &= y' - y_1. \end{aligned}$$

Again,

$$\begin{aligned} P_2 S &= M_2 S - M_2 P_2 \\ &= y' - y_2. \end{aligned}$$

Substituting in (2), we have

$$\frac{y' - y_1}{y' - y_2} = \frac{m_1}{m_2},$$

or

$$y' = \frac{m_2 y_1 - m_1 y_2}{m_2 - m_1},$$

which expresses y' in terms of known quantities.

We have now found both x' and y', and therefore solved our problem.

N. B. By the conditions of the problem the distances of P' *from* P_1 and P_2 have the ratio k. As both lines are to be measured *toward* P', they will have the *same direction when* P' *is not between* P_1 and P_2; and *opposite directions when* P' *lies between* P_1 and P_2. In the first case the ratio k must be *positive;* in the second, *negative.*

The student should carefully notice that in the first case the ratios $\frac{RP'}{SP'}$ and $\frac{P_1 R}{P_2 S}$ of equations (1) and (2) are necessarily positive; in the second, negative.

Corollary. A common case of this problem is that in which P' bisects the line P_1P_2. In this case

$$P_1P' = -P_2P',$$

and the ratio

$$k = \frac{m_1}{m_2} = -1;$$

we may therefore write

$$m_1 = -m_2;$$

and, with this simplification, the coördinates of P' become

$$x' = \frac{x_1 + x_2}{2}, \quad y' = \frac{y_1 + y_2}{2},$$

the formulas for bisection.

EXAMPLES.

(1.) On the straight line through (2, 3) and (8, – 12), find a point such that its distances from the given points respectively shall be in the ratio $-\frac{1}{2}$.

Here

$$x_1 = 2, \quad y_1 = 3, \quad x_2 = 8, \quad y_2 = -12.$$
$$m_1 = -1^*, \quad m_2 = 2.$$

Substituting these values in the expressions for x' and y', we have

$$x' = \frac{2 \times 2 - (-1)\,8}{2 - (-1)} = \frac{4+8}{3} = 4$$

$$y' = \frac{2 \times 3 - (-1)\,(-12)}{2 - (-1)} = \frac{6 - 12}{3} = -2$$

The required point is, therefore, (4, – 2).

NOTE. This problem may also be stated as follows. Find the point of trisection, nearest (2, 3), of the line which joins (2, 3) with (8, – 12).

* When k is negative, it is immaterial which of the quantities m_1 and m_2 is made negative. This will be readily seen by noticing that when the signs of both m_1 and m_2 are changed in the formulas for x' and y', the signs of the numerators and denominators are reversed, and therefore the signs of the fractions remained unchanged.

(2.) Construct the points and line given in the previous example, and verify the results by calculating the distances between the given points and the point (4, – 2).

(3.) Find the middle point of the line which joins (2, 3) and (10, 7). *Ans.* (6, 5).

(4.) Find that point of the line joining (– 4, 5) with (1, 2), whose distances from these points respectively are in the ratio 3 : 2. *Ans.* (11, – 4).

(5.) A line is drawn *from* (– 2, – 5) *to* (3, 2); to what point must it be extended, in the same direction, that its length may be doubled? to what point that it may be trebled?

Ans. (8, 9); (13, 16).

(6.) Find the middle points of the sides of the triangle mentioned in § 9, Ex. (3). *Ans.* (1, – 1), (5, 6), (0, 4).

(7.) In the same triangle, find the coördinates of the points of trisection (remote from the vertices) of the lines which join the vertices to the middle points of the opposite sides.

Ans. (2, 3).

(8.) Solve the same problem when the vertices of the triangle are (x_1, y_1), (x_2, y_2), $(x_3\ y_3)$; show that your results prove a well-known theorem of Geometry.

(9.) In any right triangle, the straight line drawn from the vertex of the right angle to the middle of the hypotenuse is equal to one half the hypotenuse. Prove this theorem by means of §§ 10, 11.

SUGGESTION. — Let the vertex of the right angle be the origin and the legs of the triangle the axes of coördinates. The vertices of the triangle may then be expressed as the points (0, 0), $(x_1, 0)$, $(0, y_2)$.

(10.) Show that the straight lines joining the middle points of the adjacent sides of any quadrilateral form a second quadrilateral whose perimeter is equal to the sum of the diagonals of the first.

CHAPTER II.

LOCUS OF AN EQUATION.

12. We have seen that the position of a point in a plane is completely determined when we know its abscissa and ordinate; i. e. when we know its distances from OY and OX. Of course, if no information is given concerning the coördinates, x and y, of a point P, the point may have any position in the plane. Now we shall often have to consider points concerning whose coördinates we know *something*, yet not enough to *determine* them. In this case we shall find that the point can no longer occupy *any* position in the plane, nor is it limited to a *single* position, but that *it is confined to a series of positions, any one of which* it may take and, *at the same time*, have its coördinates *satisfy* the conditions imposed upon them.

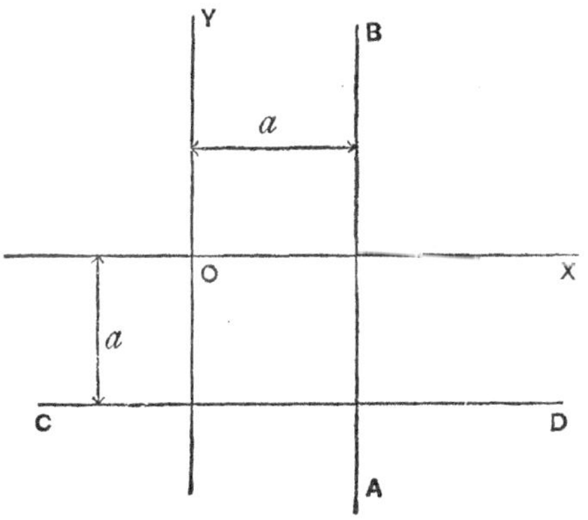

To illustrate: let us suppose that we know the abscissa of a point to be zero and have no information concerning its ordinate; i. e. we have given only $x = 0$.

This equation states that the distance of P from the axis of y is nothing; or, in other words, that P lies *on*

$O\,Y$; but as we do not know the value of y, the ordinate of P, which expresses its distance from $O\,X$, the point may lie *anywhere on* $O\,Y$. Thus we see that P may move anywhere along $O\,Y$, and yet always have the equation $x = 0$ true for it.

Likewise the single equation $y = 0$ states, in algebraic language, that the point P, of which y is the ordinate, must lie on $O\,X$, but allows it to take any position on $O\,X$.

Again: if we have given $x = a$, we know that the distance of the point P from $O\,Y$ must be a units of length; P must, therefore, lie *somewhere* on the line $A\,B$ parallel to $O\,Y$ and at the distance a from it; but it may occupy *any* position on $A\,B$.

In the same way the single equation $y = -a$ indicates that P may lie anywhere on $C\,D$, parallel to $O\,X$ and a units below it.

Let us next examine the equation $x = y$. We see at once that this equation does not *determine* x and y, the coördinates of P, but that it does show us that the coördinates of P are *always equal* to each other both in numerical value and algebraic sign; or, in other words, that the point must be just as far from one axis as from the other, and that its directions from the axes are either *both* positive or *both* negative.

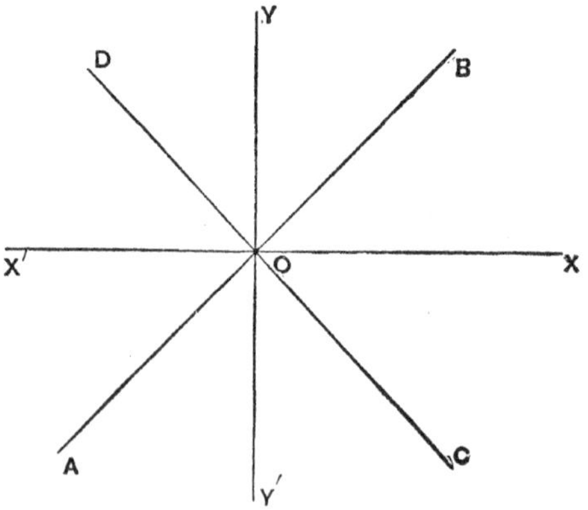

The point will satisfy both of these conditions if it is anywhere on the line $A\,B$ which bisects the angle $X\,O\,Y$, and it cannot fulfil them unless it is a point of $A\,B$. If the point lies on the part of the line in the angle $X\,O\,Y$, its coördinates are equal and positive; if on the part in the angle $X'\,O\,Y'$, they are equal and negative.

If the coördinates of P are connected by the equation $x = -y$, we readily see that P must lie somewhere on the line CD which bisects the angle YOX'.

13. The simple examples of § 12 illustrate the fact that any single equation connecting x and y, the coördinates of P, though not determining these quantities, states some law which governs their changes, and, so, confines P to some series of positions in the plane. This series of positions which a point can occupy while its coördinates satisfy a given equation is called the *locus of that equation.*

We may also define the locus of an equation as follows: —

The locus of an equation is the path of a point which moves in the plane so that at each instant its coördinates satisfy that equation when substituted therein for x and y, the variables in that equation.

14. Looking at this relation between an equation and a line from a somewhat different point of view, we notice that the line AB in the last figure has the following geometric property: —

Each point in AB is equally distant *from* OY and OX; a property which (remembering the distinction between positive and negative directions) no other point in the plane possesses. Now the distances of a point *from* OY and OX are the coördinates x and y of that point, and therefore the equation $x = y$ expresses algebraically the distinctive property of points on AB, or, as we may say, of the line AB. This equation, therefore, is called *the equation of* AB.

We may define the equation of a curve * as follows: —

The equation of a curve is the expression, in an equation, of the relation which exists between the coördinates of every point of that curve, and of no other points.

15. This relation between a curve and its equation, a relation

* The word curve is used in Analytic Geometry to indicate any geometric locus, whether a curve, a straight line, or an isolated point.

so intimate that either may be taken as a representative of the other, is the foundation of our subject, and cannot be too carefully studied.

The development of the subject will be natural and easy to the student who has so thoroughly mastered this idea that he never for an instant loses sight of it; while to one who overlooks it, or whose conception of it is vague, the whole subject must be unintelligible and difficult.

16. In § 12 we have determined the loci of several equations by observing the geometric properties which are stated in algebraic language in those equations. This method is useful for simple equations, where the geometric property is a simple one. For instance, we may obtain the locus of $x^2 + y^2 = a^2$ in this way; for, by § 10, Ex. 5, we know that $x^2 + y^2$ represents the square of the distance of the point P, which has x and y for its coördinates, from the origin. Our equation therefore states that P moves in the plane so that the square of its distance from O is constant and equal to a^2. The distance of P from O is therefore constant and equal to a, which can only be the case when P moves on the circumference of a circle with a for a radius and O for a centre. Consequently this circumference is the locus of the equation $x^2 + y^2 = a^2$.

17. There is, however, another method of determining approximately the locus of an equation, founded on the fact that the locus is the assemblage of all points whose coördinates satisfy that equation. We can often find enough of these points to show the general form of the locus; for we must bear in mind that a curve drawn at random (as in the figure) is not the locus of any equation. The equation expresses some *law* which *governs* the changes of the coördinates of P, and consequently *controls* the path of P in the plane; therefore the locus of the equation must be a curve governed by some law, and cannot

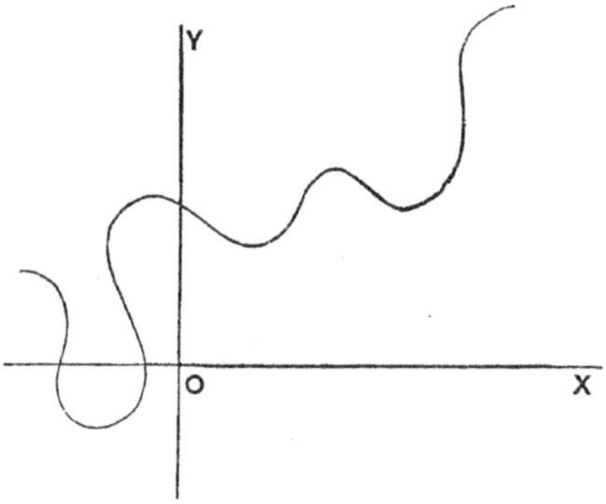

be a line drawn at random. Hence the general form of the curve will be represented by the positions of points upon it, chosen at reasonable intervals.

This method of determining a curve by points is called *plotting* the curve.

Let us *plot* the curve whose equation is $x - y + 2 = 0$.

If we choose any value for x, we can find a value for y which, together with the chosen value of x, will satisfy the equation. These values are therefore the coördinates of one point of the locus of the equation. If we choose $x = 1$, y must equal 3 in order that $x - y + 2$ may equal 0: therefore (1, 3) is a point of the required locus.

Similarly (2, 4), (4, 6), (8, 10), (0, 2), (– 1, 1), (– 2, 0), (– 3, – 1), (– 5, –3), (–8, –6), are points of the locus. Calling these points P_1, P_2, etc., to P_{10}, and constructing them as indicated by the figure, we observe that they all lie on a straight line; and from this we infer that the locus of the equation $x - y + 2 = 0$ is the straight line

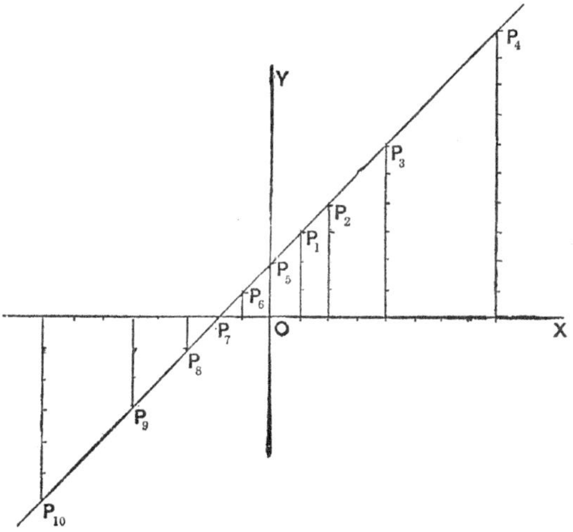

which cuts $O Y$ two units above O, and $O X$ two units to the left of O.

We shall arrive at the same result if we use the method of § 16. For the equation may be written $y = x + 2$, which interpreted in geometric language means that every point of the locus has an *ordinate two units greater than its abscissa*; or, in other words, that *the distance of each point from $O X$ is greater, by two units, than its distance from $O Y$.* Now in § 12 we showed that the straight line bisecting $X O Y$ had all its points equally distant from the two axes; the required locus is therefore parallel to that bisector, and at a distance of two units from it when measured parallel to $O Y$. It will be seen that this result agrees with the line which we have just plotted.

Let us plot the locus of $y^2 = 4x$.

In this equation:

if $x = 0, y = 0$;
if $x = 1, y = \pm 2$;
if $x = 2, y = \pm 2\sqrt{2} = \pm 2.82$;
if $x = 3, y = \pm 2\sqrt{3} = \pm 3.46$;
if $x = 4, y = \pm 4$;
if $x = 5, y = \pm 2\sqrt{5} = \pm 4.48$;
if $x = 6, y = \pm 2\sqrt{6} = \pm 4.90$;
if $x = 7, y = \pm 2\sqrt{7} = \pm 5.30$;
if $x = 8, y = \pm 4\sqrt{2} = \pm 5.64$;
if $x = 9, y = \pm 6$;

and so forth.

We must also notice that any negative value of x yields imaginary values for y; hence we conclude that there are no points of this locus with negative abscissas, i. e., no points to the left of $O Y$. Again, each abscissa has two ordinates, equal in length but opposite in direction, either of which may accompany it and satisfy the equation; from this we infer that for every point *above* OX there is a point symmetrically situated *below* OX; or we may say that the curve is symmetrical with respect to OX.

It is also evident from the equation, that the larger x is, the larger y must be; there can be no limit, therefore, to the extent of the curve to the right of OY; and it will go on forever in this direction both above and below OX, gradually receding from both axes.

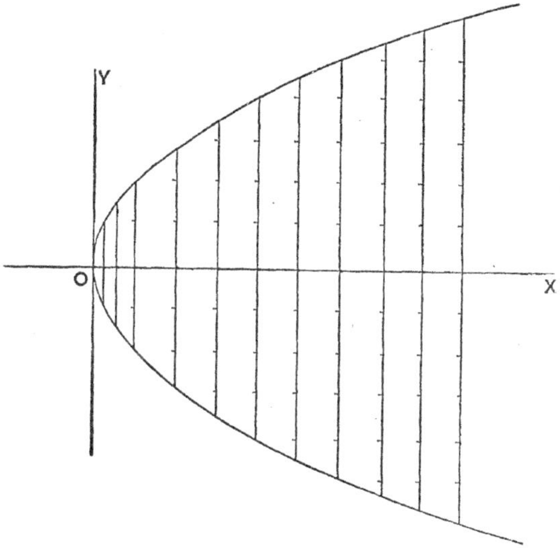

Plotting the points which we have found, we see that the form of the curve is well represented except between the points (0, 0) and (1, ± 2). It will be well, therefore, to find more points in this interval.

If $x = \frac{1}{4}$, $y = \pm 1$;

if $x = \frac{1}{2}$, $y = \pm \sqrt{2} = \pm 1.41$.

Plotting these points in addition to those already found, the curve may be represented with tolerable accuracy by a line drawn through them, and having the form indicated by these points. As we have shown, it will extend indefinitely to the right.

EXAMPLES.

Plot the curves whose equations are:—

(1.) $2x + 3y - 6 = 0$; (2.) $5x + y + 2 = 0$;

(3.) $x^2 + y^2 = 25$; (4.) $4x^2 + 9y^2 = 144$;

(5.) $4x^2 - 9y^2 = 36$; (6.) $y = x^3$;
(7.) $y^2 = x^3$; (8.) $x^2 - y^2 = 0$;
(9.) $y = \sin x$; (10.) $y = \tan x$;
(11.) $y = \log_{10} x$; (12.) $y = 2^x$.

(13.) Show that equation

$$4x^2 + 3y^2 + 2 = 0$$

has no locus.

(14.) Show that the locus of the equation

$$x^2 + 2y^2 = 0$$

consists of a single point.

18. The distance *from* the origin along the axis of x to a point where any curve cuts the axis is called an *intercept* of that curve on OX. As a curve may cut the axis of x in more than one point, it may have more than one intercept on OX; each, however, is measured from O. In the same way, a curve has one or more intercepts on OY.

The intercepts of a curve may be easily found from its equation; for it is evident, from the definitions just given, that the intercept on OX is the abscissa of the point where the curve cuts OX, and the intercept on OY, the ordinate of the point where the curve cuts OY. Let A represent the first of these points, and B the second. The coördinates of A may be represented by x_A and y_A; those of B by x_B and y_B. Now since A is a point of the curve, its coördinates must satisfy the equation of the curve, when substituted therein for x and y; moreover, $y_A = 0$, because A is a point of OX. Making the substitutions, x_A is the only unknown quantity in the equation, and can be found by solving the equation. If there are two or more values for x_A, the curve has two or more intercepts on OX, — one for each value of x_A.

Similarly the intercepts on OY may be found by substituting x_B and y_B for x and y in the equation, and then making $x_B = 0$. The resulting values of y_B are the required intercepts.

EXAMPLES.

(1.) Calculate the intercepts of the curve whose equation is

$$2x - 3y - 6 = 0 \quad \dots\dots\dots\dots (1)$$

Writing x_A for x and y_A for y, in equation (1), we have

$$2x_A - 3y_A - 6 = 0 \quad \dots\dots\dots\dots (2)$$

Substituting in (2) the value $y_A = 0$, we have

$$x_A = 3.$$

Again; to find the intercept on $O\,Y$, we have

$$2x_B - 3y_B - 6 = 0,$$

and $$x_B = 0;$$

therefore $$y_B = -2.$$

The curve therefore cuts $O\,X$ *three* units to the right of the origin and $O\,Y$ *two* units below the origin.

(2.) Find the intercepts of the curve

$$x + 3y^2 = 27.$$

The intercept on $O\,X$ may be found as in the last example, and is 27. To find the intercepts on $O\,Y$, we substitute x_B and y_B in the equation of the curve, and obtain, after making x_B equal to zero,

$$3y_B^2 = 27,$$

$$y_B^2 = 9,$$

and

$$y_B = \pm 3.$$

This curve, therefore cuts $O\,Y$ twice; once, three units above O, and again, three units below O.

NOTE. — In practice we need not substitute x_A and y_A, or x_B and y_B in the equation. We shall obtain the same result if we first put $y = 0$ in the equation, and find the corresponding values of x; then make $x = 0$ in the original equation, and find the corresponding values of y.

The results are the intercepts on $O\,X$ and $O\,Y$ respectively.

We will now seek the intercepts of the locus of

$$y^2 = 4x.$$

Making $y = 0$, we have

$$0 = 4x,$$

and

$$x = 0.$$

This shows that the intercept on OX has no length, and that the curve *passes through the origin.*

In like manner, when $x = 0$ in the equation, we have

$$y = 0.$$

This, too, shows that the curve passes through the origin; as it must cut OY at a point whose distance from O is zero.

19. We may observe that *a curve will always have its intercepts equal to zero, and therefore will pass through the origin, when every term in its equation (reduced to the simplest form) contains either x or y or both.* For, if this is the case, when we substitute for x and y in the equation the coördinates of the origin $(0, 0)$, each term vanishes and the equation is satisfied. This shows that the origin is a point of the curve.

On the other hand, if, when the equation is in its simplest form, it contains a term in which neither x nor y occurs, the curve cannot pass through the origin. For, in this case, when we substitute 0 for x and 0 for y, the equation becomes

$$C = 0,$$

where C represents a constant which is *not* 0. Therefore the equation is not satisfied by the coördinates of the origin, and the curve cannot contain that point.

EXAMPLE.

Calculate the intercepts of each of the curves whose equations are given in the examples of § 17.

20. If we wish to find the intersection of two curves, given by their equations, we have only to consider that the point of intersection of two curves is a point of *each* curve, and, therefore, that its coördinates must satisfy *each* equation, when substituted for x and y. If P' represents the required point, we shall have, after substituting its coördinates x' and y' in the equations of the two curves, two equations between the two unknown quantities x' and y', and can readily determine them by algebraic methods. If we find only one pair of values for x' and y', there is but one point of intersection. If there are several pairs of values for x' and y', there are several positions of P', i.e. several points of intersection.

We will illustrate this method by finding the point of intersection of the curves whose equations are

$$2x - y - 1 = 0 \quad \ldots\ldots\ldots\ldots \quad (1)$$

and

$$3x - y - 3 = 0 \quad \ldots\ldots\ldots\ldots \quad (2)$$

Let P' be the required point. Since P' is a point of the first curve, its coördinates satisfy equation (1).

Therefore,

$$2x' - y' - 1 = 0 \quad \ldots\ldots\ldots\ldots \quad (3)$$

Similarly, we have

$$3x' - y' - 3 = 0, \quad \ldots\ldots\ldots\ldots \quad (4)$$

because P' is a point of the locus of (2).

Eliminating between equations (3) and (4), we find

$$x' = 2, \quad y' = 3.$$

The curves, therefore, intersect in the point (2, 3).

Let the student verify this result by plotting the curves.

We will next find the intersections of the loci of

$$x - 7y + 25 = 0$$

and

$$x^2 + y^2 = 25.$$

Proceeding as before, we have two equations to determine x' and y': —

$$x' - 7y' + 25 = 0, \quad \ldots\ldots\ldots\ldots \quad (1)$$

$$x'^2 + y'^2 = 25 \quad \ldots\ldots\ldots\ldots \quad (2)$$

Eliminating between equations (1) and (2), we find that either

$$x' = 3 \text{ and } y' = 4,$$

or

$$x' = -4 \text{ and } y' = 3.$$

The curves, therefore, intersect in the two points (3, 4) and (– 4, 3); as will be seen by plotting the curves.

It will often happen that some of the values of x' and y' are imaginary. The geometrical explanation of this fact may be best illustrated by an example.

The points of intersection of the loci of

$$x + y = 4 \quad \text{and} \quad x^2 + y^2 = 4$$

will be found to have for their coördinates

$$(2 + \sqrt{-2}, 2 - \sqrt{-2}) \quad \text{and} \quad (2 - \sqrt{-2}, 2 + \sqrt{-2})$$

Since there are no real values of x' and y', we infer that there are no real points of intersection of the curves. This is readily verified by constructing the curves, which will be found to be a straight line and a circle with no points in common. Now a straight line may cut a circle in two points, and hence we always find two sets of values for x' and y' in solving such an example. These values may give two real and different points, as in the second example of this section; two real but coincident points, as is the case when the line is tangent to the circle; two imaginary points, when the line falls wholly without the circumference.*

NOTE. In practice it is not customary to substitute x' and y' for x and y in the equations, but to eliminate directly between the equations of the curves, and thus find the values of x and y which will satisfy both equations, — i. e. the coördinates of the points where the curves meet. It is safer for the begin-

* It is important for the student to understand that an "imaginary point" is not a point at all. By an imaginary point we mean nothing more nor less than a pair of imaginary values of x and y; and such a pair of values is not represented by any point whatever.

ner, however, to make the substitution; for he must always be careful to remember that x and y *in every equation are variables, and represent, in turn, the coördinates of each and every point of the locus of that equation.*

EXAMPLES.

Find the points of intersection of the curves represented by the following equations:—

(1) $2x - 2y - 11 = 0, \quad 2x - y - 7 = 0.$ *Ans.* $(\frac{3}{2}, -4)$.

(2) $x^2 + y^2 = 25, \quad 3x + 4y = 25.$ *Ans.* $(3, 4)$.

(3) $x^2 + y^2 = 25, \quad xy = 2.$

$$\text{Ans.} \begin{cases} \left(\frac{\sqrt{29}+\sqrt{21}}{2}, \frac{\sqrt{29}-\sqrt{21}}{2}\right) \\ \left(\frac{\sqrt{29}-\sqrt{21}}{2}, \frac{\sqrt{29}+\sqrt{21}}{2}\right) \\ \left(-\frac{\sqrt{29}+\sqrt{21}}{2}, -\frac{\sqrt{29}-\sqrt{21}}{2}\right) \\ \left(-\frac{\sqrt{29}-\sqrt{21}}{2}, -\frac{\sqrt{29}+\sqrt{21}}{2}\right) \end{cases}$$

(4) $y^2 = 2x, \quad x = 2.$ *Ans.* $(2, 2)$ and $(2, -2)$.

(5) $x^2 - 5x + y + 3 = 0, \quad x^2 + y^2 - 5x - 3y + 6 = 0.$

Ans. $\{(3, 3), (2, 3), (4, 1), (1, 1)\}$

(6) $\frac{x}{a} + \frac{y}{b} = 1, \quad \frac{x}{b} + \frac{y}{a} = 1.$

Ans. $\left(\frac{ab}{a+b}, \frac{ab}{a+b}\right)$

(7) The equations of the sides of a triangle are

$$x + 2y - 5 = 0,$$

$$2x + y - 7 = 0,$$

$$y - x - 1 = 0;$$

find the vertices of the triangle, and the lengths of its sides.

Ans. $\left\{ \begin{array}{ccc} (3, 1), & (2, 3), & (1, 2) \\ \sqrt{5}, & \sqrt{2}, & \sqrt{5} \end{array} \right\}$

(8) The equations of a circle and a chord of that circle are

$$x^2 + y^2 = 100 \text{ and } x + y = 14,$$

respectively; the centre of the circle is the origin: find the distance of the chord from the centre of the circle.

Ans. $7\sqrt{2}$.

CHAPTER III.

THE STRAIGHT LINE.

21. WE have learned in the preceding chapter that any equation connecting x and y has some geometric locus. Conversely, it is true that any curve which is generated by a point moving according to some law has for its equation the algebraic expression of that law.

We have seen that the loci of many equations are straight lines. Let us examine the distinctive properties of straight lines and express them in an equation, — i. e. find, in a general form, an equation which will always represent a straight line, but which may in turn represent all straight lines.

22. We will first find the equation of the straight line which cuts OX at A and OY at B. By changing the positions of A and B, this line may be made to take any desired position in the plane.

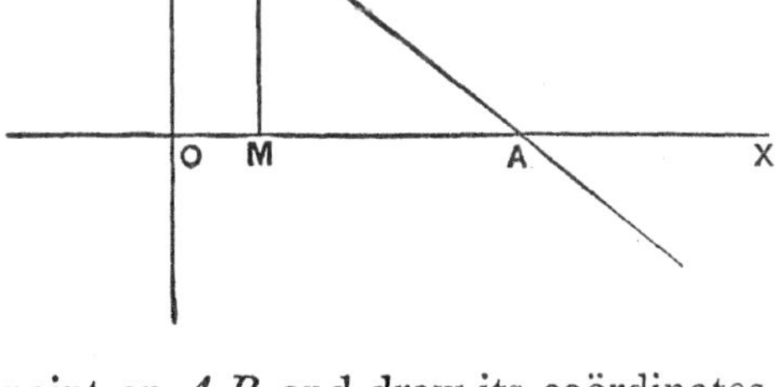

The lines OA and OB are the intercepts of AB, and will be called a and b, respectively.

Let P represent any point on AB, and draw its coördinates $OM = x$, $MP = y$.

Whatever the position of P, MP is parallel to OY, and the triangles OBA and MPA are similar. We have, therefore,

$$\frac{MP}{MA} = \frac{OB}{OA} = \frac{b}{a} \quad \ldots\ldots\ldots\ldots \quad (1)$$

But

$$MP = y,$$

and, by § 2,

$$\begin{aligned} MA &= MO + OA \\ &= OA - OM \\ &= a - x \end{aligned}$$

Substituting these values in (1), we have

$$\frac{y}{a-x} = \frac{b}{a},$$

or

$$ay = ab - bx \quad \ldots\ldots\ldots\ldots \quad (2)$$

Equation (2) may be written as follows:

$$\frac{bx}{ab} + \frac{ay}{ab} = \frac{ab}{ab},$$

or

$$\frac{x}{a} + \frac{y}{b} = 1 \quad \ldots\ldots\ldots\ldots \quad (2)$$

The last form of the equation is the one commonly used; it is called the equation of the straight line, in terms of its intercepts.

NOTE. The student should study equation (1) carefully, and convince himself that it is true, wherever P may lie on the line AB, — extended indefinitely in both directions, — but cannot be true if P moves off of that line. If, in the first case, P and B lie on the same side of OX, MP and OB have like directions and algebraic signs; the same is true of MA and OA; and, therefore, the ratios in (1) have like signs, as well as the same numerical value. If P and B lie on opposite sides of OX, MP and OB have opposite directions and algebraic signs; the same is true of $M A$ and $O A$; and, again, the ratios of (1) have like signs, together with the same numerical value. When P does not lie on $A B$, the triangles $O B A$ and $M P A$ are no longer similar, and the ratios of (1) are unequal numerically.

We see, then, that motion of P along the straight line $A B$ is consistent with equation (1), but that P is confined to positions on that line. Equation (1), therefore, expresses the distinctive properties of the straight line; and its modified forms, (2) and (3), express the invariable relation which exists between the coördinates of *any* point of $A B$. It is, then, the equation of $A B$, according to the definition of the equation of a curve, given in § 14.

EXAMPLE.

Find the equation of the straight line which makes intercepts 2 and 3 on $O X$ and $O Y$ respectively.

Here

$$a = 2 \text{ and } b = 3;$$

therefore, the required equation is

$$\frac{x}{2} + \frac{y}{3} = 1,$$

or

$$3x + 2y - 6 = 0.$$

23. In this connection it is important to notice the distinctions between absolute constants, arbitrary constants, and variables.

An *absolute constant* has a fixed value, the same wherever that constant appears. 3, -5, $\frac{1}{2}$, $\sqrt{2}$, π (the ratio of the circumference to diameter), etc., are absolute constants.

An *arbitrary constant* is a quantity which, though fixed in any particular problem, has different values in different problems of the same class; c, the circumference of a circle, and r, its radius, are examples of arbitrary constants.

A *variable* is a quantity which may have an indefinite number of values in any one connection. The length of a chord of a circle is a variable, because in the same circle it may have an indefinite number of values.

In the equation of the straight line found in § 22, we have examples of each of these three kinds of quantities: 1 is an absolute constant; a and b are arbitrary constants, fixed when the equation represents any particular line, but capable of other values as the equation represents other lines; x and y are variables, for when the equation represents any particular line (i. e. with each pair of values of a and b), these quantities are susceptible of innumerable values, as they represent, in turn, the coördinates of all points on the line.

24. While the equation of the straight line found in § 22 may represent *any* straight line* by giving appropriate values to a and b, we must often consider straight lines which are not known to us by their intercepts, but by means of other properties of the lines. In such cases, it is convenient to have the equations of the lines in terms of the determining quantities which are given. We shall, therefore, find several other forms of the equation of a straight line.

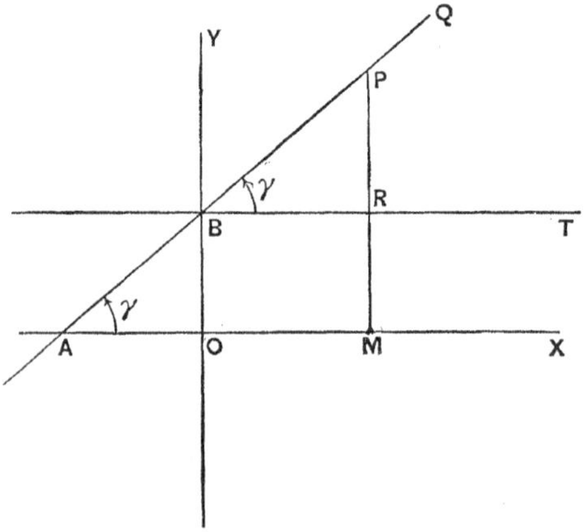

25. A straight line is determined if we know *where it cuts* OY, and also *its direction*. The first of these elements is given by the intercept on OY, b; the second is conveniently expressed by the angle which the line makes with OX, which we will call γ. The

* Provided, merely, that the line does not pass through the origin.

tangent of γ, which is called the *slope* of the line, we will indicate by λ.

In the figure, $A\,B$ represents any line; $O\,B = b$; and the angle $X\,A\,Q = \gamma$. The angle γ is always to be measured *from* $O\,X$ *to* $A\,Q$, and we usually measure it in the positive direction. P represents any point on the line, and its coördinates, $O\,M$ and $M\,P$, are x and y, respectively. Through B draw $B\,T$ parallel to $O\,X$, and in the positive direction, and let this line and $M\,P$ (or their extensions) meet in R.

From the figure and § 2, we have

$$y = M\,P = M\,R + R\,P$$
$$= O\,B + R\,P \qquad (1)$$

But

$$O\,B = b, \qquad (2)$$

and

$$T\,B\,Q = X\,A\,Q = \gamma$$

The triangle $B\,R\,P$ is a triangle of reference for γ.

From this triangle we have

$$\frac{R\,P}{B\,R} = \tan\gamma = \lambda,$$

or

$$R\,P = \overline{B\,R} \times \lambda.$$

Since

$$B\,R = O\,M = x,$$

we have

$$R\,P = \lambda\,x \qquad (3)$$

Substituting in equation (1) the values found in (2) and (3), we have

$$y = b + \lambda\,x,$$

or

$$y = \lambda\,x + b,$$

the usual form of the equation of the straight line in terms of its *intercept* and *slope*.

This is evidently the equation we seek; for it expresses, in terms of the given arbitrary constants, a relation [equation (1)] which exists between the coördinates of any point of AB, but of no other points in the plane.

REMARKS.—If γ is in the 1st or 3d quadrant, λ ($= \tan \gamma$) is positive; in this case the figure shows that RP and BR have like signs.

If γ is in the 2d or 4th quadrant, λ is negative; in this case it is easily seen that RP and BR have unlike signs.

If $\gamma = 0$ (i. e. if the line is parallel to OX), $\lambda = 0$; the equation, therefore, reduces to the form

$$y = b,$$

which we have already seen must represent a line parallel to OX (v. § 12).

If $b = 0$, the equation becomes

$$y = \lambda x,$$

which is, therefore, the equation of a straight line through the origin; this result agrees with the principle stated in § 19.

EXAMPLES.

(1.) Find the equation of the line which cuts OY at a distance of two units above the origin, and makes an angle of 45° with OX.

Here $\lambda = \tan 45^\circ = 1$ and $b = 2$; the required equation is, therefore,

$$y = x + 2.$$

This may be verified by plotting the locus.

(2.) Find the equations of the lines which cut OY three units below the origin, and make angles with OX of 135°, 30°, 120°, 210°, 330°, respectively.

(3.) Find the equations of a pair of parallel lines making an angle of 150° with OX, one passing through the origin, the other cutting OY at a distance of five units below O.

(4.) $x^2 + y^2 = 25$ is the equation of a circle with the origin as centre. Find the coördinates of the extremities of the diameter which has a slope $\frac{3}{4}$.

Ans. (4, 3) and (– 4, – 3).

26. If a straight line passes through a given point and has a given direction (indicated by its slope), it is determined in position. The fixed point we will call P_1 or (x_1, y_1), and the slope of the line, λ.

We will deduce the equation of the line in terms of these arbitrary constants, using a method which will be of great service hereafter.

The equation of § 25,

$$y = \lambda x + b, \quad \dots \dots \dots \dots \quad (1)$$

will represent the required line, if we can find such a value for b that the line with that intercept and the slope λ shall pass through P_1.

This we can easily do; for, that the locus of (1) may pass through P_1, it is necessary and sufficient that the coördinates of P_1 shall satisfy (1), when substituted therein for x and y. Making this substitution, we have

$$y_1 = \lambda x_1 + b, \quad \dots \dots \dots \dots \quad (2)$$

which may be called the *equation of condition*, that the locus of (1) shall pass through P_1, or that P_1 may lie on the locus of (1).

Now, in equation (2), x_1, y_1, and λ are the arbitrary constants which determine the position of the line whose equation we seek. We can, therefore, find b in terms of these quantities by solving equation (2). This gives

$$b = y_1 - \lambda x_1 ;$$

substituting this value of b in (1) we have

$$y = \lambda x + (y_1 - \lambda x_1),$$

which may be written

$$y - y_1 = \lambda (x - x_1).$$

This must be the equation sought, since it expresses the invariable relation between the coördinates of every point of the line (and of no other points) in terms of the required arbitrary constants.

The student should verify this equation by deducing it from a figure, without using the equations of the straight line already found. The method to be used is analogous to that of § 25.

EXAMPLES.

(1.) What is the equation of the straight line which passes through the point (2, 3) and has an inclination of 45°?

$$Ans. \begin{cases} (y-3) = (x-2), \\ \text{or } x - y + 1 = 0. \end{cases}$$

(2.) Find the equations of the straight lines which pass through (2, − 4) and make angles of 60° with OX and 30° with OY, respectively.

$$Ans. \begin{cases} y + 4 = \sqrt{3}\,(x-2), \\ y + 4 = -\sqrt{3}\,(x-2). \end{cases}$$

(3.) Find the equation of the straight line which passes through the point common to the loci of

$$2x - y - 4 = 0 \text{ and } 3x - y - 5 = 0,$$

and makes an angle 135° with OX.

$$Ans.\ x + y + 1 = 0.$$

(4.) Find the equations of a pair of parallel lines making an angle 120° with OX, and passing through the points of intersection of the loci of

$$7x + y = 50 \text{ and } x^2 + y^2 = 100.$$

$$Ans. \begin{cases} y + \sqrt{3}\,x - 8 - 6\sqrt{3} = 0, \\ y + \sqrt{3}\,x + 6 - 8\sqrt{3} = 0. \end{cases}$$

27. A straight line is also determined when it passes through two given points. Here the arbitrary constants are the coördinates of the given points, and we can find the equation of the line in terms of these quantities from the equation of the straight line found in the last section.

The equation

$$y - y_1 = \lambda (x - x_1) \quad \ldots\ldots\ldots\ldots (1)$$

represents a straight line which passes through one given point P_1, and has a slope λ. By giving to this line the proper direction or slope, we can make it pass through the second given point P_2. Let us find the appropriate value of λ.

The equation of condition which must be fulfilled in order that the locus of (1) may pass through P_2 is

$$y_2 - y_1 = \lambda (x_2 - x_1); \quad \ldots\ldots\ldots\ldots (2)$$

i.e. the coördinates of P_2 must satisfy equation (1). In equation (2), λ is the only unknown quantity; and solving for λ, we find

$$\lambda = \frac{y_2 - y_1}{x_2 - x_1} \quad \ldots\ldots\ldots\ldots (3)$$

Therefore if λ in equation (1) has this value, the locus of (1) will be a straight line through P_1 *and* P_2. Substituting this value of λ in (1), we have

$$y - y_1 = \frac{y_2 - y_1}{x_2 - x_1} (x - x_1), \quad \ldots\ldots\ldots\ldots (4)$$

which must be the required equation.

This equation is commonly written in the more symmetrical form

$$\frac{y - y_1}{y_2 - y_1} = \frac{x - x_1}{x_2 - x_1}; \quad \ldots\ldots\ldots\ldots (5)$$

but the first form is often useful.

Another convenient form of this equation, obtained by clearing of fractions and reducing, is

$$(y_1 - y_2)\, x - (x_1 - x_2)\, y + x_1 y_2 - x_2 y_1 = 0 \quad \ldots (6)$$

The student should now deduce equation (5) from a figure, without assuming any other form of the equation of the straight line.

EXAMPLES.

(1.) Find the equation of the straight line which connects the points (4, 2) and (− 3, 9).

We may choose either point as P_1, and the other will be P_2. Let

$$x_1 = 4, \quad y_1 = 2, \quad x_2 = -3, \quad y_2 = 9.$$

Substituting these values in the equation just found, we have

$$\frac{y-2}{9-2} = \frac{x-4}{-3-4},$$

or

$$\frac{y-2}{7} = \frac{x-4}{-7},$$

which may be written

$$x + y - 6 = 0,$$

the required equation in its simplest form.

(2.) Find the equations of the sides of the triangle which has for its vertices the points (2, 1), (3, − 2), (− 4, − 1).

$$\textit{Ans.} \begin{cases} 3x + y - 7 = 0, \\ x + 7y + 11 = 0, \\ x - 3y + 1 = 0. \end{cases}$$

(3.) In the same triangle, find the equations of the straight lines which connect the vertices with the middle points of the opposite sides.

$$\textit{Ans.} \begin{cases} x - y - 1 = 0. \\ x + 2y + 1 = 0. \\ x - 13y - 9 = 0. \end{cases}$$

(4.) Find the equations of the diagonals of the rectangle formed by the lines whose equations are

$$x = a, \quad x = a', \quad y = b, \quad y = b'.$$

$$\textit{Ans.} \begin{cases} (b - b')\,x - (a - a')\,y + a\,b' - a'\,b = 0, \\ (b - b')\,x + (a - a')\,y - a\,b + a'\,b' = 0. \end{cases}$$

(5.) Find the equation of the straight line which joins the

points of intersection of the two pairs of straight lines which have the equations

$$\left.\begin{array}{l} 2x + 3y - 4a = 0 \\ 2x + y - a = 0 \end{array}\right\} \text{ and } \left\{\begin{array}{l} x + 6y - 7a = 0 \\ 3x - 2y + 2a = 0. \end{array}\right.$$

Ans. $4x + 4y - 5a = 0.$

(6.) Find the equation of the straight line which joins the origin to the intersection of the loci of

$$Ax + By + C = 0 \text{ and } A'x + B'y + C' = 0.$$

Ans. $(AC' - A'C)\,x + (BC' - B'C)\,y = 0.$

(7.) Find the equations of the diagonals of the quadrilateral which has for the equations of its sides

$$3x + 4y - 13 = 0, \qquad x - 6y + 3 = 0,$$
$$2x + 3y + 6 = 0, \qquad 3x + 11y + 22 = 0.$$

Ans. $\left\{\begin{array}{l} x - y - 2 = 0, \\ 5x + 14y + 15 = 0. \end{array}\right.$

(8.) Find the equations of the medial lines of the triangle whose vertices are the points P_1, P_2, P_3.

Ans. One of the lines has the equation

$$(y_2 + y_3 - 2y_1)\,x - (x_2 + x_3 - 2x_1)\,y + (x_2y_1 - x_1y_2) + (x_3y_1 - x_1y_3) = 0;$$

the other equations may be formed from this by advancing the suffixes.

(9) Find the equations of the same lines when P_3 is the origin, and P_3P_2 is the axis of x.

Ans. $\left\{\begin{array}{l} 2y_1x + (x_2 - 2x_1)\,y - x_2y_1 = 0, \\ y_1x + (2x_2 - x_1)\,y - x_2y_1 = 0, \\ y_1x - (x_1 + x_2)\,y = 0. \end{array}\right.$

(10.) Confirm the results of Ex. (9) by means of the equations found in Ex. (8).

(11.) Prove that the medial lines of a triangle meet in a common point.

(12.) Prove that the diagonals of a parallelogram bisect each other.

SUGGESTION. Take one side of the parallelogram for the axis of x, and one extremity of that side as the origin.

(13.) Prove that the lines which join the middle points of the opposite sides of a trapezoid bisect each other.

(14.) If E and F are the middle points of the opposite sides, $A\,D$, $B\,C$, of a parallelogram $A\,B\,C\,D$, the straight lines $B\,E$, $D\,F$, trisect the diagonal $A\,C$. Prove this theorem (v. §§ 10, 11, 20, 27).

(15.) Prove that (2, 3), (−4, −1), (8, 7) lie on one straight line.

28. A straight line is determined in position if we know the length of a perpendicular drawn to it *from* the origin, and the direction of this perpendicular (represented by the angle which it makes with $O\,X$).

We will next find the equation of the straight line in terms of these arbitrary constants.

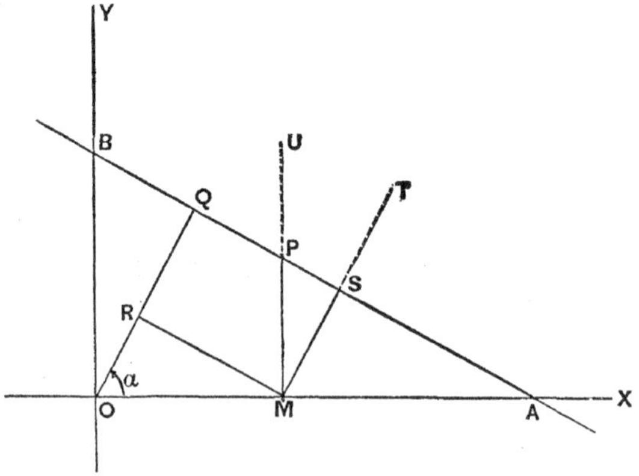

Let $A\,B$ represent any line in the plane, $O\,Q$ the perpendicular upon it from O, and $X\,O\,Q$ the angle which determines the direction of $O\,Q$. $O\,Q$ is measured *from* O; and the angle is measured *from* $O\,X$ to $O\,Q$. We will call the angle α, and the perpendicular p; it will be seen that p will always be positive, for the positive direction of the terminal line of an angle is *from the vertex along the boundary of the angle.*

Let P be any point of AB, with coördinates $x = OM$, $y = MP$. From M draw perpendiculars to OQ and AB meeting them in R and S respectively.

From the figure and § 3 we have

$$\begin{aligned} QOY &= QOX + XOY \\ &= XOY - XOQ \\ &= 90^\circ - \alpha. \end{aligned}$$

If we always take T and U on MS and MP respectively, so that MT shall have the same direction as its parallel OQ, and MU as OY, we may write

$$TMU = QOY = 90^\circ - \alpha.$$

Now, ORM is always a triangle of reference for the angle QOX, or $(-\alpha)$; and MSP is a triangle of reference for the angle TMU, or $(90^\circ - \alpha)$.

From the figure we have, by § 2,

$$\begin{aligned} p = OQ &= OR + RQ \\ &= OR + MS \quad \ldots\ldots\ldots\ldots \quad (1) \end{aligned}$$

Now

$$\frac{OR}{OM} = \cos(-\alpha) = \cos\alpha;$$

therefore,

$$\begin{aligned} OR &= \overline{OM} \times \cos\alpha \\ &= x\cos\alpha \quad \ldots\ldots\ldots\ldots\ldots \quad (2) \end{aligned}$$

Again,

$$\frac{MS}{MP} = \cos(90^\circ - \alpha) = \sin\alpha;$$

therefore,

$$\begin{aligned} MS &= \overline{MP} \times \sin\alpha \\ &= y\sin\alpha \quad \ldots\ldots\ldots\ldots\ldots \quad (3) \end{aligned}$$

Substituting in equation (1) the values found in (2) and (3), we have

$$p = x\cos\alpha + y\sin\alpha,$$

or $$x \cos \alpha + y \sin \alpha = p,$$

the usual form of the equation.

Since this equation expresses the necessary relation between the coördinates of every point of the straight line $A\,B$ (and of no other point), it is the equation of that line.

Let the student obtain this form of the equation of the straight line from the equation in terms of the intercepts,

$$\frac{x}{a} + \frac{y}{b} = 1.$$

This may be done by expressing a and b in terms of p and α.

NOTE. — It is important to notice that the equation which we have deduced in this section will remain the equation of $A\,B$, if we consider $Q\,O$ the positive direction of the perpendicular, instead of $O\,Q$. For the angle α is measured from $O\,X$ to the *positive direction* of the perpendicular, and $(90° - \alpha)$ is measured from the *positive direction* of the perpendicular to $O\,Y$.

Therefore, by reversing the direction of the perpendicular, we change each angle by $180°$, and only change the algebraic signs of their cosines. Each term in the equation, therefore, has its sign changed, and the equation is still true for all points on $A\,B$, and for no other points.

EXAMPLES.

(1.) Find the equation of the straight line such that the perpendicular upon it from the origin bisects the angle $X\,O\,Y$, and equals $\sqrt{\tfrac{1}{2}}$.

Here

$$p = \sqrt{\tfrac{1}{2}} \text{ and } \alpha = 45°;$$

therefore

$$\cos \alpha = \sqrt{\tfrac{1}{2}} \text{ and } \sin \alpha = \sqrt{\tfrac{1}{2}}.$$

The required equation is, therefore,

$$x\sqrt{\tfrac{1}{2}} + y\sqrt{\tfrac{1}{2}} = \sqrt{\tfrac{1}{2}},$$

or $$x + y = 1.$$

(2.) A circle, with a radius 4, has the origin for its centre. Find the equations of the tangents to the circle at the extremities of the diameter which makes an angle 120° with $O X$.

$$Ans. \begin{cases} x - \sqrt{3}\, y + 8 = 0. \\ x - \sqrt{3}\, y - 8 = 0. \end{cases}$$

29. We have found, in §§ 22–28, the equation of a straight line in different forms; and we have in several cases shown how one of these forms is merely a modification of others.

It is noticeable that in each form the equation is of the *first degree.* We will now examine the general form of the equation of the first degree between x and y, and show that it always expresses a law governing these quantities such that the point (x, y) will be confined to motion on a straight line.

The general form of the equation of the first degree may be written:—

$$A x + B y + C = 0; \quad \ldots \ldots \ldots \quad (1)$$

where A represents the sum of the coefficients of all the terms which contain x; B, of all which contain y; and C, the sum of all terms which contain neither x nor y.

If we solve equation (1) for y, we have

$$y = -\frac{A}{B} x - \frac{C}{B}, \quad \ldots \ldots \ldots \quad (2)$$

which is in the form

$$y = \lambda x + b, \quad \ldots \ldots \ldots \quad (3)$$

$$\left(\text{where } \lambda = -\frac{A}{B} \text{ and } b = -\frac{C}{B}\right)$$

which is known to be the equation of a straight line.

Now, whatever the values of A, B, and C, $-\frac{A}{B}$ is the slope of some line. For the slope of a line is the tangent of the angle which the line makes with $O X$; and as this angle changes from 0° to 180°, its tangent passes through all possible values from ∞

to $-\infty$. There is some angle, therefore, between 0° and 180°, which has its tangent equal to $-\frac{A}{B}$, and a straight line having this inclination has $-\frac{A}{B}$ for its slope.

Again, $-\frac{C}{B}$ is the intercept of some line on OY; for a line may cut OY anywhere, and so have any intercept.

If, then, we form the equation of the straight line which has the slope $-\frac{A}{B}$, and $-\frac{C}{B}$ for its intercept on OY, the equation must be

$$y = \left(-\frac{A}{B}\right)x + \left(-\frac{C}{B}\right)$$

or

$$Ax + By + C = 0.$$

Therefore, whatever the values of A, B, and C, equation (1) has for its locus some straight line. But (1) represents in turn all equations of the first degree between x and y. Therefore, all such equations are equations of straight lines.

From the preceding discussion, we see that the values of the slope and intercept on OY can be easily obtained from the equation of the line. For, if this equation is put in the form

$$Ax + By + C = 0,$$

we have

$$\lambda = -\frac{A}{B},\ b = -\frac{C}{B}.$$

NOTE. The student should notice that the principle of this section will be proved if we can put the general equation in any of the forms which we know to represent straight lines.

EXAMPLES.

(1) Put the general equation of the first degree in the form

$$\frac{x}{a} + \frac{y}{b} = 1,$$

and obtain values for a and b in terms of the coefficients A, B, and C.

$$Ans.\ a = -\frac{C}{A},\quad b = -\frac{C}{B}$$

(2) Calculate the intercepts and slope of the straight line whose equation is

$$2x + 3y - 6 = 0.$$

Here

$$A = 2,\quad B = 3,\quad C = -6,$$

and we easily find

$$a = 3,\quad b = 2,\quad \lambda = -\frac{2}{3}.$$

(3) Find the intercepts and slope for each of the lines which have the following equations: —

$$4x - 3y - 12 = 0,$$
$$x + y = 3,$$
$$x \cos\alpha + y \sin\alpha = p.$$

30. It is obvious that two equations, of which the second can be obtained from the first by multiplication by a constant factor, will have precisely the same locus, since any pair of coördinates which satisfies one of these equations also satisfies the other. Thus the two equations,

$$2x - 3y - 6 = 0,$$

$$6x - 9y - 18 = 0,$$

both represent the line whose intercepts are 3 and -2.

No less important, though perhaps less obvious, is the following fact

If two equations of the first degree have the same locus, the second can be obtained from the first by multiplication by a constant factor.

For, let

$$Ax + By + C = 0, \quad \ldots\ldots\ldots\ldots \quad (1)$$

$$A_1x + B_1y + C_1 = 0, \quad \ldots\ldots\ldots\ldots \quad (2)$$

be the two equations of the first degree which are supposed to have the same locus. This locus, as we know, is a straight line whose slope λ and intercept b may be computed from the first equation by the formulæ

$$\lambda = -\frac{A}{B}, \qquad b = -\frac{C}{B};$$

from the second equation by the formulæ

$$\lambda = -\frac{A_1}{B_1}, \qquad b = -\frac{C_1}{B_1}.$$

When we remember that the two equations are supposed to represent one and the same line, we see that the two values of λ just obtained must be equal; and the same is true of the two values of b. Therefore

$$-\frac{A}{B} = -\frac{A_1}{B_1}, \qquad -\frac{C}{B} = -\frac{C_1}{B_1};$$

or

$$\frac{A_1}{A} = \frac{B_1}{B} = \frac{C_1}{C} \quad \ldots\ldots\ldots\ldots \quad (3)$$

Let us now multiply (1) by the constant $\frac{A_1}{A}$. This gives

$$\frac{A_1}{A} A x + \frac{A_1}{A} B y + \frac{A_1}{A} C = 0,$$

or, when we replace $\frac{A_1}{A}$ in the second term by its value $\frac{B_1}{B}$, and in the third term by its value $\frac{C_1}{C}$ [v. (3)],

$$\frac{A_1}{A} A x + \frac{B_1}{B} B y + \frac{C_1}{C} C = 0,$$

and this reduces at once to equation (2). Thus we see that (2) may be obtained by multiplying (1) by the constant $\frac{A_1}{A}$.

31. Let us reduce the general equation of the first degree to the form found in § 28.

Whatever the locus of

$$A\,x + B\,y + C = 0, \quad \ldots\ldots\ldots\ldots (1)$$

its equation may be expressed in the form

$$x \cos \alpha + y \sin \alpha - p = 0; \quad \ldots\ldots\ldots (2)$$

for every straight line can have its equation written in any of the forms found in §§ 22–28.

Regarding (1) and (2) as equations of the same straight line, their first members must either be identical, or one must be some multiple of the other (v. § 30). Let the unknown multiplier, which will make (1) identical with (2), be represented by R.

Then,

$$R\,A\,x + R\,B\,y + R\,C = 0, \quad \ldots\ldots\ldots (3)$$

and

$$x \cos \alpha + y \sin \alpha - p = 0 \quad \ldots\ldots\ldots (4)$$

are identical. Therefore

$$R\,A = \cos \alpha, \quad R\,B = \sin \alpha, \quad R\,C = -p.$$

We have from Trigonometry

$$\cos^2 \alpha + \sin^2 \alpha = 1\,;$$

substituting the values of cos α and sin α,

$$R^2\,A^2 + R^2\,B^2 = 1\,;$$

and solving for R^2,

$$R^2 = \frac{1}{A^2 + B^2},$$

and

$$R = \frac{1}{\pm\sqrt{A^2 + B^2}}.$$

Using this value of R, we have

$$\cos a = \frac{A}{\pm\sqrt{A^2 + B^2}},$$

$$\sin a = \frac{B}{\pm\sqrt{A^2 + B^2}},$$

$$p = -\frac{C}{\pm\sqrt{A^2 + B^2}}.$$

We may use either sign with the radical, but must, of course, use the same one in the values of $\cos a$, $\sin a$, and p. We have decided to take p positive; therefore, we will in each case choose that sign with the radical which will make $-\frac{C}{\sqrt{A^2 + B^2}}$ positive. This must be the opposite sign to that of C.

NOTE. — The reason why we may choose either sign with $\sqrt{A^2 + B^2}$ (i. e. have two multipliers, R), is found in the note to § 28. It is there shown that the same line will have two equations in the form

$$x \cos a + y \sin a = p,$$

if we allow either algebraic sign with p; and that the values of $\cos a$, $\sin a$, and p, will be the same numerically, but have opposite signs in the two equations.

Now, if we choose that sign for $\sqrt{A^2 + B^2}$ which makes p positive, we get certain values for $\cos a$ and $\sin a$; these values will only have their signs reversed when we change the sign of $\sqrt{A^2 + B^2}$, so that the second value of a differs from the first by $180°$, as was shown in the note to § 28.

EXAMPLES.

(1.) Reduce the equation

$$4x - 3y - 7 = 0$$

to the form

$$x \cos a + y \sin a = p.$$

Here

$$A = 4, \; B = -3, \; C = -7;$$

therefore

$$R = \frac{1}{+\sqrt{A^2 + B^2}} = \frac{1}{\sqrt{16 + 9}} = \frac{1}{5}.$$

We choose the positive sign with the radical, because the sign of C is negative.

$$\frac{4}{5}x - \frac{3}{5}y = \frac{7}{5}$$

is, therefore, the required equation, and

$$\cos \alpha = \frac{4}{5}, \; \sin \alpha = -\frac{3}{5}, \; p = \frac{7}{5}.$$

(2.) Find the distance from the origin to the line which has the equation

$$2\sqrt{3}\,x - 2y + 7 = 0,$$

and the angle which this perpendicular makes with OX.

Ans. $\frac{7}{4}$; 150°.

(3.) Find the length of the perpendicular from the origin on the line whose equation is

$$a(x - a) + b(y - b) = 0.$$

Ans. $\sqrt{a^2 + b^2}$.

32. In the preceding sections we have found those forms of the equation of the straight line which will be most useful hereafter. We have also shown that any equation of the first degree between x and y must represent a straight line.

We will now make use of the equations which we have found to solve several problems relating to straight lines and points, the results of which are convenient for reference.

33. Let us find the angle between two straight lines given by their equations: —

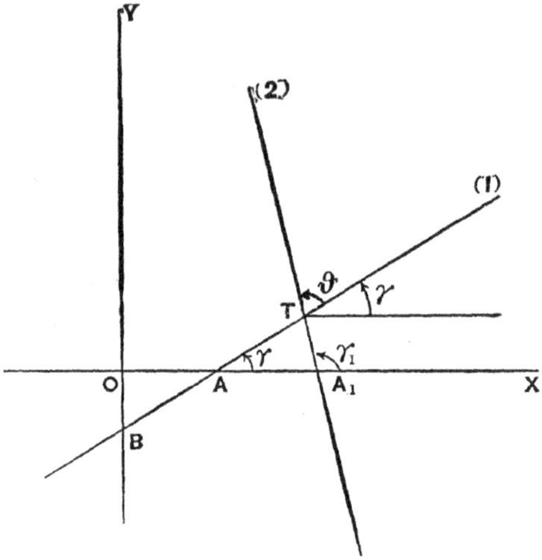

First, suppose the equations of the lines to be in terms of intercept and slope. We may write the equations:

$$y = \lambda x + b, \quad . \ . \quad (1)$$
$$y = \lambda_1 x + b_1 \quad . \ . \quad (2)$$

Here, $\lambda = \tan \gamma$, and $\lambda_1 = \tan \gamma_1$, γ and γ_1 being the angles which the lines (1) and (2) respectively make with $O\,X$, as represented in the figure. Let ϑ be the angle which the line (2) makes with the line (1); i. e. ϑ is measured *from* (1) *to* (2). This is the angle we seek.

By § 3,
$$\gamma + \vartheta = \gamma_1, \quad \ldots \ldots \ldots \ldots \quad (3)$$

as is evident when we draw through T, the point of intersection of the lines, a line parallel to $O\,X$.

From (3), we have
$$\vartheta = \gamma_1 - \gamma.$$

Therefore,
$$\tan \vartheta = \frac{\tan \gamma_1 - \tan \gamma}{1 + \tan \gamma_1 \tan \gamma}.$$

Substituting the values of $\tan \gamma_1$ and $\tan \gamma$, we have
$$\tan \vartheta = \frac{\lambda_1 - \lambda}{1 + \lambda_1 \lambda} \quad \ldots \ldots \ldots \quad (4)$$

This gives $\tan \vartheta$ in terms of known quantities, and ϑ can be easily found from its tangent by Trigonometry.

If the equations of the lines are

$$A\,x + B\,y + C = 0, \quad \dots\dots\dots\dots \quad (1')$$

$$A_1x + B_1y + C_1 = 0, \quad \dots\dots\dots\dots \quad (2')$$

we have only to find the values of λ and λ_1 in terms of A, B, A_1, B_1, and substitute the values thus found in (4).

By § 29,

$$\lambda = -\frac{A}{B} \text{ and } \lambda_1 = -\frac{A_1}{B_1}.$$

Therefore,

$$\tan\vartheta = \frac{-\frac{A_1}{B_1} - \left(-\frac{A}{B}\right)}{1 + \left(-\frac{A_1}{B_1}\right)\left(-\frac{A}{B}\right)},$$

or

$$\tan\vartheta = \frac{A\,B_1 - A_1\,B}{A\,A_1 + B\,B_1} \quad \dots\dots\dots\dots \quad (5)$$

NOTE. — It should be carefully remembered that ϑ is measured from (1) to (2), and so will be positive or negative according as γ_1 is greater or less than γ. By interchanging the lines (1) and (2), therefore, we shall change the sign of ϑ, and so change the sign of its tangent. The latter fact is also evident from the values of tan ϑ, in either of which the proposed interchange will reverse the sign of the fraction.

Corollary 1. If the lines (1) and (2) are parallel, the angle ϑ is 0°, or 180°, and

$$\tan\vartheta = 0.$$

That this may be true, we must have the numerators of the fractions in (4) and (5) equal to 0.

The conditions for parallel lines are, therefore,

$$\lambda_1 - \lambda = 0, \text{ or } \lambda_1 = \lambda;$$

or,

$$A\,B_1 - A_1B = 0, \text{ or } A\,B_1 = A_1B.$$

Corollary 2. If the lines are mutually perpendicular, ϑ is 90°, or 270°, and

$$\tan \vartheta = \infty .$$

Making the denominators of the fractions in (4) and (5) equal to 0, we have for the conditions of perpendicular lines

$$1 + \lambda_1 \lambda = 0, \text{ or } \lambda_1 = -\frac{1}{\lambda};$$

or,

$$A A_1 + B B_1 = 0, \text{ or } A A_1 = - B B_1.$$

We may state the results of these corollaries, thus:

Parallel lines have *the same slope.*

Lines at right angles to each other must have *the slope of one equal to the negative of the reciprocal of the slope of the other.*

EXAMPLES.

(1.) Find the angle which the line

$$3x + y - 7 = 0$$

makes with the line

$$y = 4x + 2.$$

Here, by the statement of the question, the angle is to be measured *from* the last line. We have, then,

$$\lambda = 4, \quad \lambda_1 = -3,$$

and

$$\tan \vartheta = \frac{-3 - 4}{1 + (-3)\,4} = \frac{7}{11};$$

therefore

$$\vartheta = \tan^{-1} \frac{7}{11}$$

If we use the other formula we have

$$A = 4, \quad B = -1, \quad A_1 = 3, \quad B_1 = 1,$$

and

$$\tan \vartheta = \frac{4 \times 1 - 3 \times (-1)}{4 \times 3 + (-1) \times 1} = \frac{7}{11},$$

which agrees with the previous result.

(2.) Find the angle which the line

$$3x + y = 7$$

makes with the line

$$2x - y = 3.$$

Ans. $\frac{\pi}{4}$.

(3.) Find the angle which the last line mentioned in Ex. (2) makes with

$$3x + y = 5.$$

(4.) Find the angles of the quadrilateral mentioned in § 27, Ex. 7.

Ans. $\left\{\tan^{-1} -\frac{22}{21},\quad \tan^{-1}\frac{15}{16},\quad \tan^{-1} -\frac{1}{3},\quad \tan^{-1}\frac{21}{53}\right\}$

(5.) Find the equation of the straight line which cuts OY four units below the origin, and is parallel to the line

$$2x - 5y + 4 = 0;$$

find the equation of a line which has the same intercept on OY, and is perpendicular to the line

$$\frac{x}{5} - \frac{y}{2} = 1.$$

Ans. $\begin{cases} 2x - 5y - 20 = 0, \\ 5x + 2y + 8 = 0. \end{cases}$

(6.) Recollecting that the angle between two lines is equal to the angle between the perpendiculars upon those lines from the origin, find an expression for the tangent of the angle which the line

$$x \cos \alpha_1 + y \sin \alpha_1 = p_1$$

makes with the line

$$x \cos \alpha + y \sin \alpha = p.$$

(7.) By the same method, deduce the formula already found for the tangent of the angle which the line

$$A_1 x + B_1 y + C_1 = 0$$

makes with the line

$$Ax + By + C = 0.$$

(8.) Find the angle between the lines

$$4x - 3y + 5 = 0,$$

and

$$6x + 8y + 17 = 0.$$

(9.) Show that the diagonals of a square are perpendicular to each other.

(10.) Show that the second quadrilateral mentioned in § 11, Ex. 10, is a parallelogram.

34. It often happens that we have a line

$$Ax + By + C = 0, \quad \ldots \ldots \ldots \ldots \quad (1)$$

and wish to find the equation of another line parallel or perpendicular to it. The following considerations are very useful in such cases.

The slope of any line parallel to (1) is $-\frac{A}{B}$ (§ 33, Cor. 1). Accordingly the equation of such a line may be written (§ 25)

$$y = -\frac{A}{B}x + b,$$

or

$$Ax + By - bB = 0.$$

The equation of any line parallel to (1) *may be so written that its equation differs from* (1) *only in the value of the constant term.*

Similarly the equation of any line perpendicular to (1) may be written (§ 33, Cor. 2)

$$y = \frac{B}{A}x + b,$$

or

$$Bx - Ay + bA = 0.$$

The equation of any line perpendicular to (1) *may be obtained by interchanging the coefficients of x and y in* (1) *and changing the sign of one of them, and then suitably altering the constant term.*

These two rules enable us to write down the equation of the desired line completely, except for the constant term. The fact that the constant term is not determined by these rules should not surprise us, for the mere statement that the desired line is to be parallel or perpendicular to (1) does not determine the line completely. Some further statement with regard to the position of the line must be made before the constant term can be determined. This may be done, for instance, as follows:

Let us consider the problem: To find the equation of the line which passes through the point (x_1, y_1) and is parallel to (1).

Since this line is to be parallel to (1), its equation can, by the first of the above rules, be written in the form

$$Ax + By = k, \quad \ldots\ldots\ldots\ldots \quad (2)$$

where, for convenience, we have written the constant term k on the right-hand side of the equation. The question which remains to be settled is: What is the value of the constant k? Since the line (2) is to pass through (x_1, y_1), the following equation of condition must be satisfied:

$$Ax_1 + By_1 = k,$$

and it will be seen that this equation gives the value of the unknown constant k in terms of the known constants A, B, x_1, y_1. Substituting this value of k in (2), we find as the equation of the desired line

$$Ax + By = Ax_1 + By_1.$$

By precisely the same method we find as the equation of the line which passes through the point (x_1, y_1) and is perpendicular to (1)

$$Bx - Ay = Bx_1 - Ay_1.$$

EXAMPLES.

(1.) Find the equations of two straight lines through $(2, -1)$, parallel and perpendicular, respectively to the line

$$5x + 3y + 7 = 0.$$

(2.) Find the equations of two lines through the origin, parallel and perpendicular, respectively, to the line

$$3x + 5y - 4 = 0.$$

(3.) Find the equation of the line which passes through the point common to the lines

$$x - 2y - a = 0,$$

and $$x + 3y - 2a = 0,$$

and is parallel to the line

$$3x + 4y = 0.$$

Ans. $3x + 4y - 5a = 0.$

(4.) Find the equations of the perpendiculars from the vertices to the opposite sides of the triangle given in § 20, Ex. 7.

$$Ans. \begin{cases} x + y - 4 = 0 \\ 2x - y - 1 = 0 \\ x - 2y + 3 = 0. \end{cases}$$

(5.) Through the vertices of this triangle lines are drawn parallel to the opposite sides; find the equations of the sides of the triangle thus formed.

$$Ans. \begin{cases} x - y - 2 = 0 \\ x + 2y - 8 = 0 \\ 2x + y - 4 = 0. \end{cases}$$

(6.) Find the condition that the lines

$$x + (a + b)y + c = 0,$$

and $$a(x + ay) + b(x - by) + d = 0$$

may be parallel; that they may be mutually perpendicular.

Ans. $b = 0, \quad b^2 - a^2 = 1.$

(7.) Choosing the axes as in § 27, Ex. 9, prove that the perpendiculars erected at the middle points of the sides of a triangle meet in the point $\left(\frac{x_2}{2}, \frac{x_1^2 + y_1^2 - x_1 x_2}{2 y_1}\right)$.

(8.) In a similar manner, show that the perpendiculars from the vertices to the opposite sides of a triangle meet in the point $\left(x_1, \frac{x_1 x_2 - x_1^2}{y_1}\right)$.

(9.) Prove that the common intersection of the medial lines of a triangle, the intersection of the perpendiculars at the middle points of its sides, and the intersection of the perpendiculars from the vertices to the opposite sides lie on one straight line.

(10.) Show that the first of these points is one of the points of trisection of the line which joins the other two. To which of these points is it nearer? (v. § 11.)

35. Let us find the distance of a given point P_1 *from* a given straight line.

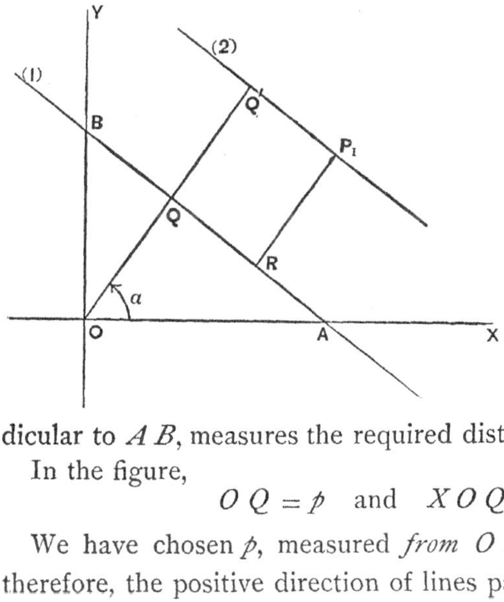

Let AB be the given line, and let its equation be

$$x \cos a + y \sin a = p \quad (1)$$

Draw a line, (2), through P_1, parallel to (1); and from O draw a perpendicular to the lines (1) and (2), meeting them in Q and Q', respectively. RP_1, perpendicular to AB, measures the required distance of P_1 *from* AB.

In the figure,

$$OQ = p \quad \text{and} \quad XOQ = a.$$

We have chosen p, measured *from* O *to* Q, positive; this is, therefore, the positive direction of lines parallel to OQ.

Since (1) and (2) are parallel lines, the perpendiculars upon them, from O, make the same angle, a, with OX; the equation of (2) is, therefore,

$$x \cos a + y \sin a = p', \quad \ldots\ldots\ldots\ldots \quad (2)$$

where $p' = OQ'$.

From the figure and § 2, we have

$$RP_1 = QQ' = OQ' - OQ$$
$$= p' - p \quad \ldots\ldots\ldots\ldots \quad (3)$$

Now, P_1 is a point of the line (2); therefore its coördinates must satisfy equation (2), and we have

$$x_1 \cos \alpha + y_1 \sin \alpha = p'.$$

Substituting this value of p' in equation (3), we have

$$R P_1 = x_1 \cos \alpha + y_1 \sin \alpha - p, \quad \ldots \ldots \quad (4)$$

which expresses the required distance in terms of known quantities.

NOTE. We see from the figure, that $R P_1$ is positive or negative according as P_1 and O are on opposite sides of $A B$ or on the same side. The same result is obtained from equation (4). For, in the first case, p' is greater than p, and both are positive. In the second, when P_1 is nearer $A B$ than O is, p' is positive and less than p; when P_1 is farther from $A B$ than O is, p' will itself be negative, according to the principle explained in § 28, Note.

36. We may solve the problem of the last section when the given line has the equation

$$A x + B y + C = 0 \quad \ldots \ldots \ldots \ldots \quad (1)$$

by reducing equation (1) to the form

$$x \cos \alpha + y \sin \alpha = p,$$

as explained in § 31.

The value of the required distance is, therefore,

$$x_1 \frac{A}{\sqrt{A^2 + B^2}} + y_1 \frac{B}{\sqrt{A^2 + B^2}} - \left(- \frac{C}{\sqrt{A^2 + B^2}}\right),$$

or

$$\frac{A x_1 + B y_1 + C}{\sqrt{A^2 + B^2}}, \quad \ldots \ldots \ldots \ldots \quad (2)$$

where the radical has the opposite sign to that of C in order to make

$$p \left[= - \frac{C}{\sqrt{A^2 + B^2}} \right] \text{ positive.}$$

EXAMPLES.

(1.) How far is the point (3, – 2) from the line

$$3x - 4y + 2 = 0?$$

The formula of § 36 shows that the required distance is

$$\frac{3 \times 3 - 4(-2) + 2}{-\sqrt{3^2 + (-4)^2}} = -3\tfrac{4}{5}.$$

The negative sign indicates that the point lies on the same side of the line with the origin.

(2.) Calculate the distance of the point (7, 4) from the line

$$8x + 6y = 45.$$

Ans. $3\frac{1}{2}$; and the line lies between the point and the origin.

(3.) Find the distances of the origin from the lines

$$x \cos a + y \sin a = p,$$
$$Ax + By + C = 0.$$

Ans. $-p, -\dfrac{C}{\sqrt{A^2 + B^2}}$.

(4.) In the triangle given in § 20, Ex. 7, find the lengths of the perpendiculars from the vertices to the opposite sides.

(5.) Find the lengths of similar lines in the triangle given in § 9, Ex. 3, and show by your results that the origin lies within the triangle.

(6.) We may obtain the distance between a given point and a given line by finding, first, the equation of the perpendicular from the point to the line; then, the point where this perpendicular meets the given line; and, lastly, the distance between this point and the given point.

By this method, calculate the distance of the point (– 8, 3) from the line

$$3x - 2y + 4 = 0.$$

Confirm your result by applying the formula of § 36.

(7.) Using the same method, show that the distance between P_1 and the line

$$A\,x + B\,y + C = 0$$

is

$$\frac{A\,x_1 + B\,y_1 + C}{\sqrt{A^2 + B^2}}.$$

(8.) If the given point lies on the given line, its distance from the line is nothing. Show that the formulas of §§ 35, 36, give the same result.

(9.) Two parallel lines are drawn at an inclination ϑ to the axis of x: one through the point (a, b), the other through (a', b'). Show that the distance between these lines is

$$(a' - a)\sin\vartheta - (b' - b)\cos\vartheta.$$

37. The area of a triangle can be found when the coördinates of its vertices are given. For we have only to calculate the length of one side, by § 10, and the perpendicular upon this side from the opposite vertex, by § 36. The half-product of these quantities is the measure of the area. Let P_1, P_2, P_3, be the vertices of the triangle. The equation of the line P_2P_3 (i. e. of the indefinite line determined by P_2 and P_3) is, by § 27,

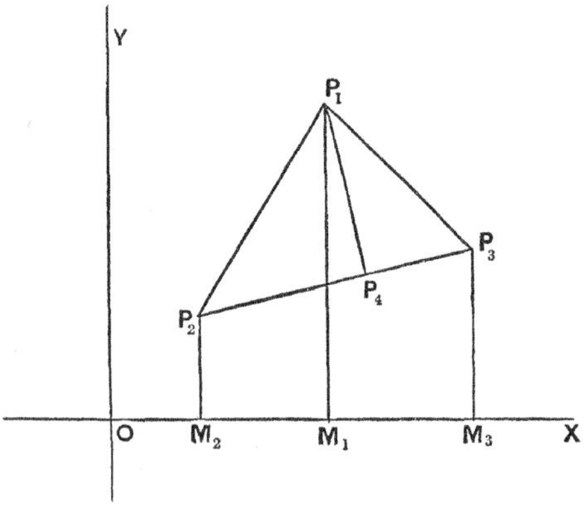

$$(y_2 - y_3)\,x - (x_2 - x_3)\,y + x_2 y_3 - x_3 y_2 = 0 \quad . \; . \; . \; . \quad (1)$$

The distance of P_1 from this line is, by § 36,

$$P_4 P_1 = \frac{(y_2 - y_3)\,x_1 - (x_2 - x_3)\,y_1 + x_2 y_3 - x_3 y_2}{\sqrt{(y_2 - y_3)^2 + (x_2 - x_3)^2}}. \quad . \; . \quad (2)$$

By § 10, we have

$$P_2 P_3 = \sqrt{(y_2 - y_3)^2 + (x_2 - x_3)^2}. \quad \ldots \ldots (3)$$

If A represents the area of the triangle,

$$2\,A = \overline{P_4\,P_1} \times \overline{P_2\,P_3}$$

$$= (y_2 - y_3)\,x_1 - (x_2 - x_3)\,y_1 + x_2 y_3 - x_3\,y_2 \quad \ldots \; (4)$$

Equation (4) may be written

$$2\,A = (x_1 - x_2)\,y_3 + (x_2 - x_3)\,y_1 + (x_3 - x_1)\,y_2.$$

This form is evidently symmetrical with respect to the coördinates of P_1, P_2, and P_3, as should be the case; for it is, of course, immaterial which side is used as the base of the triangle, when calculating the area.

NOTE. As only the *length* of $P_4\,P_1$ is needed in calculating the area of the triangle, we need only notice the numerical value of our result. The algebraic sign will generally differ when we take different sides for the base.

EXAMPLES.

(1.) Find the area of the triangle which has for its vertices the points (2, 3), (– 1, 4) (6, – 5).

Calling the points P_1, P_2, and P_3, respectively, we have

$$2\,A = [2 - (-1)]\,(-5) + (-1 - 6)\,3 + (6 - 2)\,4$$

$$= -15 - 21 + 16 = -20.$$

Therefore the area is 10.

The unit of area is, of course, the square which has each side equal to the unit of length in terms of which the coördinates of the given points are expressed.

(2.) What is the area of the triangle given in § 20, Ex. 7.

(3.) Find the area of the quadrilateral given in § 27, Ex. 7.

Ans. $\frac{57}{2}$.

(4.) Obtain the formula for the area of the triangle $P_1\,P_2\,P_3$ [§ 37, Figure] by subtracting the trapezoid $M_2\,P_2\,P_3\,M_3$ from the sum of the trapezoids $M_2\,P_2\,P_1 M_1$ and $M_1\,P_1\,P_3\,M_3$.

(5.) Prove that the three straight lines joining the middle points of the sides of a triangle divide the triangle into four equal triangles.

38. In all our work hitherto, we have used the rectangular system of coördinates. It is often more convenient to denote the position of a point by means of its *direction and distance from a fixed point* in the plane, than by means of its distances from two fixed lines.

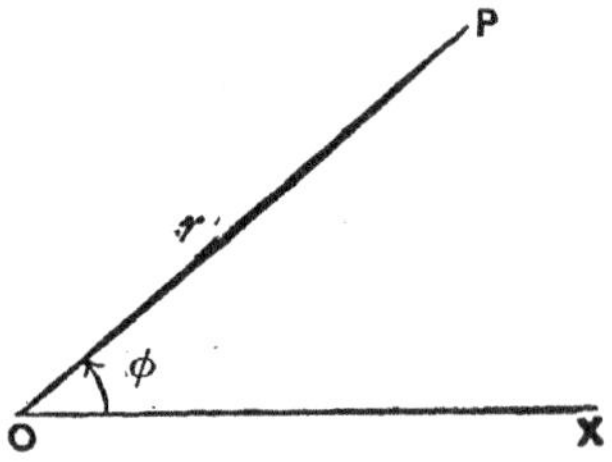

If OX in the figure is a fixed line, and O a fixed point of that line, the position of P is completely given by the angle $XOP = \phi$, and the distance $OP = r$.

The two elements, or *coördinates,* which determine the position of the point P are, therefore, r, called the *radius vector*, and ϕ, called the *vectorial angle*; ϕ is measured *from* the fixed straight line, called the *initial line*, and r *from* O, called the *pole.*

This system of coördinates is called *polar.*

It will be readily seen that the simple equation

$$r = a$$

has for its locus the circumference of a circle with O, the pole, for its centre, and a for its radius. This example is sufficient to show the student that certain problems may be conveniently treated by this system of coördinates.

39. We will find the equation of the straight line, using the polar system.

In the figure on the next page, O is the pole, OX the initial line, and AB any straight line in the plane.

The position of the line is determined by p, equal to OQ, the perpendicular *from* the pole *to* the line, and α, equal to XOQ, the angle which p makes with the initial line.

Let P be *any* point on AB, with coördinates

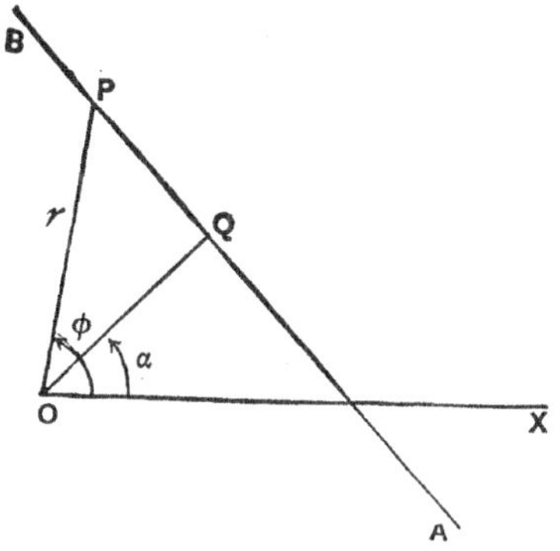

$$r = OP, \quad \phi = XOP.$$

Now from the figure and § 3 we may write

$$\begin{aligned} QOP &= QOX + XOP \\ &= XOP - XOQ \\ &= \phi - \alpha. \quad \ldots \ldots (1) \end{aligned}$$

Again; the triangle QOP must be a triangle of reference for the angle QOP, and therefore we may write

$$\frac{OQ}{OP} = \cos QOP = \cos(\phi - \alpha) \quad \ldots \ldots \ldots (2)$$

Substituting in (2) the values of OQ and OP, and clearing of fractions, we have

$$r \cos(\phi - \alpha) = p,$$

which must be the required equation, since it expresses a necessary relation between r and ϕ, the coördinates of *any* point on AB (and of no other points), in terms of the chosen arbitrary constants, p and α.

CHAPTER IV.

TRANSFORMATION OF COÖRDINATES.

40. WHEN the equation of a curve referred to one system of coördinates is known, it is often desirable to obtain the equation of the same curve referred to a different system. We shall be able to find formulas which connect the coördinates of *any* point in the plane in the *first* system with those of the *same* point in the *second*, and thus render easy the desired change.

41. We will first compare the coördinates of a point referred to one set of axes with the coördinates of the *same* point referred to a *parallel* set.

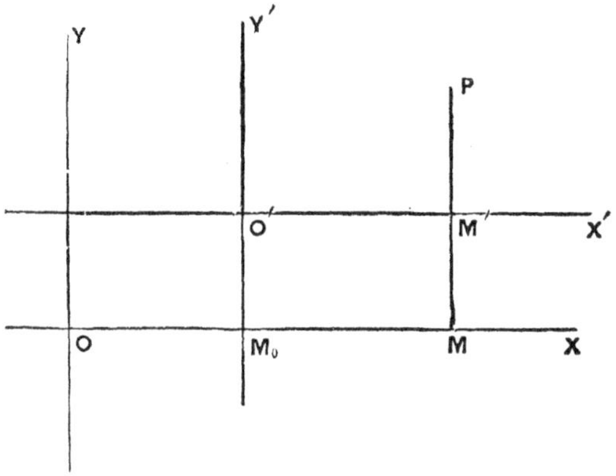

Let OX and OY be the original set of axes; $O'X'$ and $O'Y'$ the new set, parallel respectively to the old.

O', the new origin, has coördinates in the old system which we will call x_0 and y_0.

Then,

$$x_0 = OM_0, \quad y_0 = M_0O'.$$

Let P represent any point in the plane; its coördinates referred to XOY are

$$x = OM, \quad y = MP.$$

Referred to $X' O' Y'$, the coördinates of P are $O' M'$ and $M' P$; these we will call x' and y' respectively, to distinguish them from x and y; but we must remember that x' and y' are here *variables* in the *same sense* that x and y are variables.

From the figure and § 2 we have

$$OM = OM_0 + M_0 M = OM_0 + O' M' \quad . \; . \; . \; . \; (1)$$

$$MP = MM' + M' P = M_0 O' + M' P. \quad . \; . \; . \; . \; (2)$$

Substituting in these equations the values of the lines as given above, we have

$$\left. \begin{array}{l} x = x_0 + x' \\ y = y_0 + y' \end{array} \right\} \quad . \; . \; . \; . \; . \; . \; . \; . \; . \; . \; . \; (3)$$

which are evidently the equations sought, since they enable us to find values for either pair of coördinates for P, in terms of the other pair and the constants x_0 and y_0 which determine the relative positions of the two sets of axes.

We can best illustrate the use of these formulas by an example.

The equation of a line referred to rectangular axes is

$$2x + 3y = 5;$$

find the equation of the same line, referred to a parallel system which has the point (2, 4) for its origin.

Here

$$x_0 = 2 \text{ and } y_0 = 4.$$

The equations for transformation are, therefore,

$$x = 2 + x', \quad y = 4 + y'.$$

These equations mean that *any* point in the plane has its abscissa in the *first system* greater than its abscissa in the *second* by *two* units of length, and that the *old ordinate* exceeds the *new* by *four* units.

Now the locus of the given equation comprises all the points which have twice their *old abscissas* plus thrice their *old ordinates* equal to five; consequently it consists of those points which have twice *the new abscissa increased by two* plus thrice *the new ordinate increased by four* equal to five.

Its equation in the new system is, therefore,

$$2(x' + 2) + 3(y' + 4) = 5,$$

or

$$2x' + 4 + 3y' + 12 = 5,$$

or

$$2x' + 3y' + 11 = 0.$$

Since in this equation the accents are only used to distinguish the new coördinates from the old, now that we have changed to the new system and there is no longer opportunity for confusion, we will use the usual symbols for the coördinates of the point which generates the locus, and write the equation

$$2x + 3y + 11 = 0.$$

The student should construct the two sets of axes, and then plot the loci of the two equations, and show that these loci are one and the same straight line.

NOTE. — Since the results of this section have been obtained by the principle of § 2, and without reference to the angle XOY, the same formulas may be used for transforming from one *oblique* system to a *parallel oblique* system. [v. § 8.]

42. We will next find the formulas by which we may change from one rectangular system to another, *keeping the same origin.*

Let OX and OY be the original axes; OX' and OY', the new axes. Let the angle XOX', which the new axis of x makes with the old, be called θ.

The coördinates of P (any point in the plane) in the old system are $x = OM$ and $y = MP$; in the new system they are $x' = OM'$ and $y' = M'P$. From M' draw $M'R$ perpen-

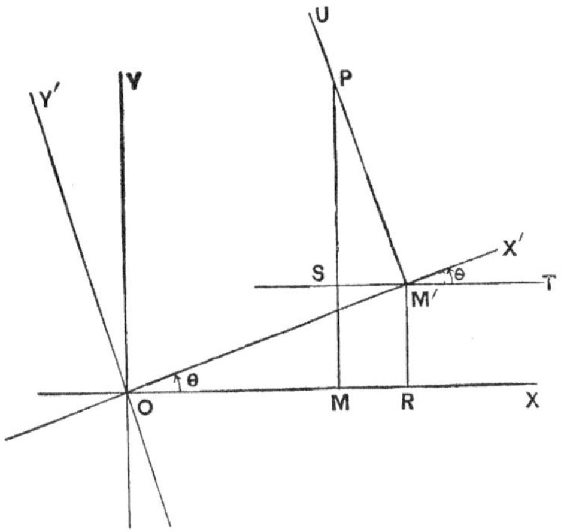

dicular to OX, and $M'T$ parallel to OX. Let the latter line (or its extension) meet MP (or its extension) in S. Take U on the line $M'P$ so that the angle $X'M'U$ shall be 90° of positive rotation. T and X' are, of course, to be chosen so that $M'T$ and $M'X'$ shall have the directions of OX and OX'.

From the figure and § 3, we have

$$\begin{aligned} TM'U &= TM'X' + X'M'U \\ &= \vartheta + 90^\circ. \end{aligned}$$

The triangles ORM' and $M'SP$ are therefore triangles of reference for ϑ and $(90^\circ + \vartheta)$, respectively.

Now, from the figure and § 2, we have

$$x = OM = OR + RM = OR + M'S \quad \ldots \quad (1)$$

$$y = MP = MS + SP = RM' + SP \quad \ldots \quad (2)$$

But

$$OR = OM' \cos\vartheta = x' \cos\vartheta \quad \ldots \quad (3)$$

$$M'S = M'P \cos(90^\circ + \vartheta) = -M'P \sin\vartheta = -y' \sin\vartheta \quad (4)$$

$$RM' = OM' \sin\vartheta = x' \sin\vartheta \quad \ldots \quad (5)$$

$$SP = M'P \sin(90^\circ + \vartheta) = M'P \cos\vartheta = y' \cos\vartheta \quad \ldots \quad (6)$$

Substituting in equations (1) and (2) the values found in (3)–(6), we have

$$\left.\begin{aligned} x &= x' \cos\vartheta - y' \sin\vartheta \\ y &= x' \sin\vartheta + y' \cos\vartheta \end{aligned}\right\} \quad \ldots \quad (7)$$

which are the required equations of transformation, since they connect the old and new coördinates by means of the arbitrary constant ϑ which determines the relative positions of the two sets of axes.

EXAMPLE.

The equation of a curve is

$$x^2 + y^2 = a^2; \quad \ldots \ldots \quad (1)$$

find the equation of the same curve when referred to axes which make an angle of 30° with the original axes, but retain the same origin.

Here

$$\vartheta = 30°; \quad \sin\vartheta = \frac{1}{2}; \quad \cos\vartheta = \frac{\sqrt{3}}{2}.$$

The formulas for transformation are, therefore,

$$x = \frac{\sqrt{3}}{2}x' - \frac{1}{2}y' \quad \text{and} \quad y = \frac{1}{2}x' + \frac{\sqrt{3}}{2}y'.$$

Since the coördinates of any point of the curve, when referred to the old system, must satisfy equation (1), their equivalents in terms of the new coördinates of those points must satisfy equation (1); therefore we have

$$\left(\frac{\sqrt{3}\,x' - y'}{2}\right)^2 + \left(\frac{x' + \sqrt{3}\,y'}{2}\right)^2 = a^2, \quad \ldots \quad (2)$$

or

$$\frac{3\,x'^2 - 2\sqrt{3}\,x'y' + y'^2}{4} + \frac{x'^2 + 2\sqrt{3}\,x'y' + 3\,y'^2}{4} = a^2,$$

which reduces to the form

$$x'^2 + y'^2 = a^2. \quad \ldots \ldots \quad (3)$$

If we drop the accents — because we have now left the old system and need no longer fear a confusion of the old and new coördinates of any point — we may write equation (3)

$$x^2 + y^2 = a^2. \quad \ldots \ldots \quad (4)$$

Since (4) is identical in form with (1), we see that the curve has the same law governing the coördinates of its points when referred to the new axes as when referred to the old. This is evidently true from the geometrical properties of the curve; for the locus of (1) is a circle with its centre at the origin. [v. § 16.]

Corollary. — The formulas of § 42 become much simplified when $\vartheta = 90°$. In this case $\sin\vartheta = 1$ and $\cos\vartheta = 0$: the equations for transformation therefore become

$$x = -y' \quad \text{and} \quad y = x'.$$

We may obtain the same result by considering that in this case we merely interchange the axes of x and y; at the same time *changing the positive direction* of the horizontal axis.

43. To change from a rectangular to a polar system, with *the origin for pole*, and *the axis of x for initial line*.

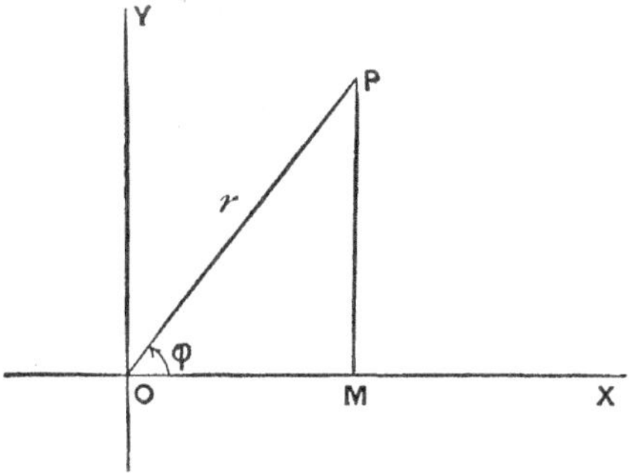

Let P be any point in the plane. Its coördinates in the rectangular system are

$$x = OM, \; y = MP;$$

in the polar system, they are

$$r = OP, \; \phi = XOP;$$

OMP is a triangle of reference for ϕ; therefore we have

$$OM = OP \times \cos\phi, \text{ or } x = r\cos\phi, \quad \ldots \quad (1)$$

$$MP = OP \times \sin\phi, \text{ or } y = r\sin\phi, \quad \ldots \quad (2)$$

which are the required equations for transformation, because they connect the rectangular and polar coördinates of any point in the plane.

EXAMPLE.

Find the polar equation of the circle

$$x^2 + y^2 = a^2,$$

referred to the centre as pole.

We have seen in the example of § 42 that the position of the axis of x does not change the equation of this curve, provided the centre is origin. Therefore, if the centre is the pole, whatever the direction of the initial line, the formulas for transformation are

$$x = r \cos \phi, \; y = r \sin \phi.$$

Substituting for x and y, in the equation of the circle, their values in terms r and ϕ, we have

$$r^2 \cos^2 \phi + r^2 \sin^2 \phi = a^2,$$

or

$$r^2 (\cos^2 \phi + \sin^2 \phi) = a^2,$$

or

$$r^2 = a^2;$$

which may be written

$$r = \pm a,$$

a result which agrees with the polar equation of the circle found in § 38.

The double sign is explained by the fact that, with any value of ϕ, the radius vector, measured in either the positive or the negative direction, meets the curve at a distance a from the pole.

44. In §§ 41–43, we have obtained the equations for transforming coördinates in three simple cases. By combining these formulas, however, we can change from a rectangular system to any other rectangular or polar system, or *vice versa;* for we can first change the origin to any desired point, by § 41; then, by § 42, make the axes take any direction; and lastly, if desired, change to polar coördinates, by § 43.

There is an important general principle concerning the transformation of coördinates which we will now establish.

Suppose the equation of a certain curve is an algebraic equation of the nth degree in the rectangular coördinates x, y; and suppose we transform to a new system of rectangular coördinates x', y' situated in any manner whatever. The equation of the curve may, and usually will, take on a form very different from what it had at first, but *it will still be an algebraic equation of the nth degree in the coördinates* x', y'. In order to prove that this statement is correct, let us consider how it is we pass from the original form of the equation of the curve to the new form. For this purpose we must first apply formulæ (3), § 41, in order to bring the origin to the desired position and then formulæ (7), § 42. In each of these formulæ the old coördinates are expressed as polynomials of the first degree in terms of the new. These transformations will, therefore, leave the equation in the form of an algebraic equation of degree no higher than n. It is not entirely obvious, however, that the degree of the new equation might not in some cases be less than n through the cancellation of the terms of highest degree. We have, therefore, as yet merely shown that the degree of an algebraic equation cannot be *raised* by a change from one system of rectangular coördinates to another. Suppose, now, that the degree of the equation could be lowered. Then starting with the equation in x', y', let us transform back to the original system of coördinates x, y. This transformation must evidently bring us back precisely to the equation we started with, so that if the degree of the equation had been lowered when we passed from the first system of coördinates to the second, it would be raised when we pass back from the second system to the first; and this is impossible, since, as we just saw, no transformation of coördinates can raise the degree of an equation.

We see, therefore, that the degree of the equation of a curve depends on the curve itself, and not at all on the special system of rectangular coördinates used. This fact justifies us in classifying curves according to the degree of their equations, and, in fact, it is customary to speak of curves of the first degree, of the second degree, etc. We have already seen that the only curve of the first degree is the straight line. It is with the study of curves of the second degree that we shall be concerned from now on.

Besides *algebraic* curves, which we classify according to their degree, there are also *transcendental* curves whose equation in rectangular coördinates cannot be written in algebraic form. Examples of such curves will be found in Ex. (9)–(12), p. 19.

The following examples may be solved by the application of one or more of the methods of transformation explained in §§ 41–43.

EXAMPLES.

(1.) The equation of a curve is

$$x^2 + y^2 - 4x - 6y = 18;$$

find the equation of the same curve, referred to axes, parallel to the old, through the point (2, 3).

Ans. $x^2 + y^2 = 31$.

(2.) The equation of a curve is

$$y^2 - x^2 = 6;$$

what is its equation when referred to lines which bisect the angles between the old axes?

Ans. $xy = 3$.

(3.) When we change from one rectangular system to another without changing the origin, $x^2 + y^2$ must equal $x'^2 + y'^2$, beause each denotes the square of the distance of the same point om the origin. Verify this by squaring and adding the expresons for x and y, in terms of x' and y', given in § 42.

(4.) The equation of a curve is

$$4xy - 3x^2 = a^2;$$

rn the axes through an angle whose tangent is 2, and find the ew equation of the curve.

Ans. $x^2 - 4y^2 = a^2$.

(5.) Find the polar equation of the line

$$Ax + By + C = 0,$$

hen the origin is the pole of the new system, and the axis of x e initial line.

Ans. $(A \cos \phi + B \sin \phi) r + C = 0$.

(6.) Under similar circumstances, find the polar equation of e curve which has for its rectangular equation

$$\frac{x^2}{a^2} + \frac{y^2}{b^2} = 1.$$

Ans. $r^2 = \frac{a^2 b^2}{a^2 \sin^2 \phi + b^2 \cos^2 \phi}$.

(7.) The equation of a curve is

$$2x^2 + 3y^2 - 16x + 18y + 53 = 0;$$

nd its polar equation, when $(4, -3)$ is the pole, and the initial ne is parallel to the axis of x.

Ans. $r^2 = \frac{6}{3 \sin^2 \phi + 2 \cos^2 \phi}$.

(8.) By combining the results of §§ 41, 42, obtain formulas or changing, at once, from one rectangular system to any other ectangular system.

CHAPTER V.

THE CIRCLE.

45. A *circle* may be defined as the path of a point which moves in a plane so that its distance from a fixed point in that plane is constant.

Let us find the equation of the circle ; i. e. an equation whose locus *must be* a circle, but which may in turn represent *all* circles.

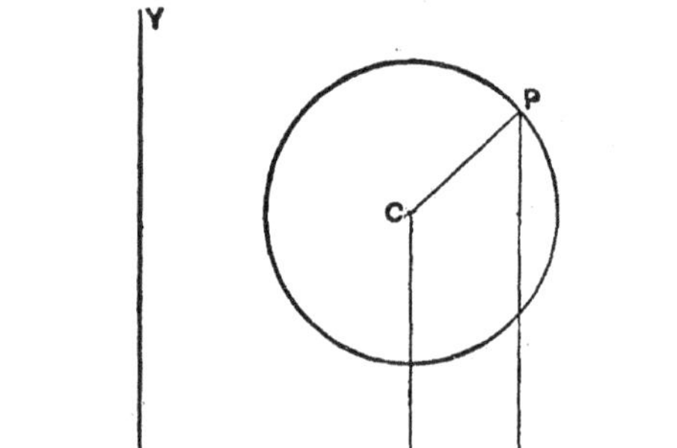

In the figure, C is the fixed point, with coördinates $O\ M_c = x_c$ and $M_c\ C = y_c$; P, with coördinates $O\ M = x$ and $M\ P = y$, is the point which generates the curve.

Letting a represent the constant distance of P from C, we may express the definition of the circle by the equation

$$C\,P = a. \qquad (1)$$

By § 10, the distance between the two points C and P may be expressed in terms of their coördinates, and we have

$$C\,P = \sqrt{(x - x_c)^2 + (y - y_c)^2}.$$

Substituting this value in (1), we have

$$\sqrt{(x - x_c)^2 + (y - y_c)^2} = a,$$

and clearing of radicals,

$$(x - x_c)^2 + (y - y_c)^2 = a^2, \quad . \; . \; . \; . \; . \; . \; (2)$$

which must be the equation of the circle, for it expresses in an equation, connecting x and y, the essential property of the circle generated by the point (x, y).

Corollary 1. When the centre is at the origin

$$x_c = 0 \text{ and } y_c = 0;$$

the equation becomes in this case

$$x^2 + y^2 = a^2,$$

a result already obtained in § 16.

Corollary 2. When a diameter is taken as the axis of x, and its left-hand extremity as the origin,

$$x_c = a \text{ and } y_c = 0;$$

in this case the equation of the circle is

$$(x - a)^2 + y^2 = a^2,$$

or,

$$x^2 + y^2 = 2\,a\,x.$$

EXAMPLES.

(1.) Write the equation of the circle whose radius is 5, and which has for its centre the point (3, 4); reduce the equation, and show from its form that the circle passes through the origin.

(2.) A circle, with radius 4, has its centre on $O\,X$ and is tangent to $O\,Y$: what is its equation if it lies on the right of $O\,Y$? What if on the left?

$$Ans. \begin{cases} x^2 + y^2 - 8\,x = 0, \\ x^2 + y^2 + 8\,x = 0. \end{cases}$$

(3.) Obtain the equation of Cor. 1, from the first form of the.

equation of the circle, by transforming the axes to a parallel set through the point (x_c, y_c).

(4.) Obtain the equation of Cor. 2 from each of the preceding forms, by transformation of coördinates.

(5.) Prove analytically that any angle inscribed in a semicircle is a right angle.

SUGGESTION.—Take the diameter bounding the semicircle as axis of x.

(6.) Find the equations of the three circles whose diameters are the sides of the triangle formed by the lines

$$x + 2y - 5 = 0,$$
$$2x + y - 7 = 0,$$
$$x - y + 1 = 0.$$

$$Ans. \begin{cases} x^2 + y^2 - 5x - 4y + 9 = 0, \\ x^2 + y^2 - 3x - 5y + 8 = 0, \\ x^2 + y^2 - 4x - 3y + 5 = 0. \end{cases}$$

(7.) Find the equation of the circle whose centre is the origin, and which is tangent to the line

$$2x - y + 3 = 0.$$

Ans. $5(x^2 + y^2) = 9.$

(8.) The equation of a chord of the circle

$$x^2 + y^2 = 100$$

is

$$7x + y = 50;$$

find the equation of the circle which has this chord for a diameter.

(9.) Find the points of meeting of the circles

$$x^2 + y^2 - 2x - 4y - 20 = 0,$$
$$x^2 + y^2 - 14x - 16y + 100 = 0.$$

Ans. (4, 6), (5, 5).

(10.) Find the equation of the chord common to the circles given in the last example, and show that this chord is perpendicular to the line joining their centres.

(11.) Show that the circles

$$x^2 + y^2 - 4x - 6y - 23 = 0,$$
$$x^2 + y^2 - 24x - 6y + 137 = 0,$$

are tangent at the point (8, 3).

(12.) Prove that the perpendicular upon a diameter of a circle from any point of the circumference is a mean proportional between the segments of the diameter.

(13.) Prove that if two circumferences are tangent internally, and the radius of the larger is the diameter of the smaller, any chord of the larger drawn from the point of contact is bisected by the circumference of the smaller.

SUGGESTION. — Take the point of contact for the origin, and a diameter for the axis of x; the equation of any chord may then be written

$$y = \lambda x.$$

(14.) If two straight lines are drawn through the point of contact of two circles, the chords of the intercepted arcs are parallel. Prove when the circles are tangent externally, taking the origin at the point of contact.

46. Equation (2) when expanded becomes

$$x^2 + y^2 - 2x_c x - 2y_c y + (x_c^2 + y_c^2 - a^2) = 0.$$

This equation has the form

$$x^2 + y^2 + ax + by + c = 0, \quad . \quad . \quad . \quad . \quad . \quad . \quad . \quad . \quad (1)$$

where a, b, c are constants. The circle is therefore a curve of the second degree.

Conversely any equation of the form (1) can be shown to represent a circle, provided its locus is a curve of any sort.*

* The reason for this restriction will be apparent presently.

Before proving this fact we will illustrate it by a numerical example. Consider the equation

$$x^2 + y^2 - 4x + 4y - 1 = 0.$$

This equation is of the form (1), and we will prove that it represents a circle, and at the same time find the centre and radius of this circle by the following device. We arrange the terms of the equation in three groups. First, the two terms involving x, then two terms involving y, and then, transposed to the right-hand side, the constant term; thus:

$$x^2 - 4x \qquad + y^2 + 4y \qquad = 1.$$

Now apply to the first group of terms the process, familiar from algebra, of completing the square by adding the square of half the coefficient of x. In the same way complete the square for the second group of terms by adding the square of half the coefficient of y. The two quantities which we have just added to the left-hand side of the equation must, of course, also be added on the right, and we thus get

$$(x - 2)^2 + (y + 2)^2 = 9.$$

This, however, is [§ 45, (2)] the equation of the circle with centre at (2, − 2) and radius 3.

Precisely this method of completing the square, which we have here applied in a numerical case, can be applied without change to the general equation (1), and gives

$$\left(x + \frac{a}{2}\right)^2 + \left(y + \frac{a}{2}\right)^2 = -c + \frac{a^2}{4} + \frac{b^2}{4}. \quad . \quad . \quad (2)$$

Provided the right-hand side of the equation is positive, we have here the equation of the circle with centre at $\left(-\frac{a}{2}, -\frac{b}{2}\right)$ and radius $\sqrt{-c + \frac{a^2}{4} + \frac{b^2}{4}}$. If, however, the right-hand side of (2) is negative, this reasoning is no longer possible, since the radius we should thus obtain is an imaginary quantity. In fact in this case equation (2) has no locus, i.e. there is no real point where coördinates satisfy this equation. For the left-hand side of (2), being the sum of two squares, is essentially positive, while the right-hand side is now supposed negative.

There remains to be considered the case in which the right-hand side of (2) is zero. In this case the circle consists of a single point $\left(-\frac{a}{2}, -\frac{b}{2}\right)$, since the left-hand side of (2), being the sum of two squares, can vanish only when each of these squares vanishes. This point is commonly regarded as a circle of radius zero.

Summing up, we may say that equation (1) represents a circle, a point, or has no real locus according as the right-hand side of (2) is positive, zero, or negative. It is usual to designate these three cases by saying we have respectively a *real*, *null*, or *imaginary* circle.

The general equation of the second degree in x and y is much more general than equation (1), for not only may the terms in x^2 and y^2 have coefficients different from unity, but there may be a term involving the product xy. Thus the general equation of the second degree may be written

$$Ax^2 + Bxy + Cy^2 + Dx + Ey + F = 0. \quad . \quad . \quad . \quad (3)$$

This equation usually represents a curve which is much more complicated than a circle. It will represent a circle (real, null, or imaginary) if $B = 0$ and $A = C \neq 0$, for then (3) has the form

$$Ax^2 + Ay^2 + Dx + Ey + F = 0,$$

and this equation reduces to (1) if we divide it through by A

EXAMPLES.

Find the centres and radii of the following circles:

(1) $x^2 + y^2 + 6x - 4y + 9 = 0$. *Ans.* $(-3, 2)$, 2.

(2) $x^2 + y^2 - 6x - 10y + 34 = 0$. *Ans.* $(3, 5)$, 0.

(3) $x^2 + y^2 - 3x + y + 2 = 0$. *Ans.* $\left(\frac{3}{2}, -\frac{1}{2}\right)$, $\frac{1}{2}\sqrt{2}$.

(4) $12x^2 + 12y^2 + 16x - 12y + 3 = 0$. *Ans.* $\left(-\frac{2}{3}, \frac{1}{2}\right)$, $\frac{2}{3}$.

47. We know from Geometry that a circle is determined by any three points on the circumference. We may, therefore, obtain the equation of a circle which passes through *any three given points.*

Let P_1, P_2, and P_3 be the three given points.

Since equation (2), § 45, represents in turn all circles, it will represent the required circle when we give appropriate values to the arbitrary constants, x_c, y_c, and a, in that equation.

That the locus of

$$(x - x_c)^2 + (y - y_c)^2 = a^2 \quad \text{. . . . . (1)}$$

may pass through P_1, P_2, and P_3, it is necessary and sufficient that the coördinates of these points should satisfy equation (1); these conditions are expressed by the three equations:—

$$(x_1 - x_c)^2 + (y_1 - y_c)^2 = a^2,$$
$$(x_2 - x_c)^2 + (y_2 - y_c)^2 = a^2,$$
$$(x_3 - x_c)^2 + (y_3 - y_c)^2 = a^2.$$

In these equations the arbitrary constants, x_c, y_c, and a, are the quantities whose values we seek, while the other quantities are the known coördinates of the given points. We have, then, three equations between three unknown quantities, and can determine them all by elimination.

Substituting in (1) the values thus found for x_c, y_c, and a, we have the equation of the required circle.

EXAMPLES.

(1.) Find the equation of the circle which passes through the points (9, 6), (10, 5), (3, – 2).

Calling these points P_1, P_2, and P_3, the three equations from which we may obtain the proper values of x_c, y_c, and a are

$$(9 - x_c)^2 + (6 - y_c)^2 = a^2 \quad \ldots\ldots\ldots (1)$$

$$(10 - x_c)^2 + (5 - y_c)^2 = a^2 \quad \ldots\ldots\ldots (2)$$

$$(3 - x_c)^2 + (-2 - y_c)^2 = a^2 \quad \ldots\ldots\ldots (3)$$

We can easily eliminate a from these equations, and obtain two new equations connecting x_c and y_c, from which we shall find

$$x_c = 6,\ y_c = 2.$$

Substituting these values in either of the equations (1), (2), (3), we obtain

$$a^2 = 25.$$

We have now determined all the constants in the required equation, which must be

$$(x - 6)^2 + (y - 2)^2 = 25,$$

or

$$x^2 + y^2 - 12x - 4y + 15 = 0.$$

(2.) Find the circle through (2, – 3), (3, – 4), (– 2, – 1).

Ans. $x^2 + y^2 + 8x + 20y + 31 = 0.$

(3.) Find the equation of the circle circumscribed about the triangle given in § 20, Ex. 7.

Ans. $3x^2 + 3y^2 - 13x - 11y + 20 = 0.$

(4.) Solve the last three problems by starting from equation (1) of § 46.

48. The term *secant* is used in Analytic Geometry to denote a straight line which cuts a curve in two or more points.

If one of the points in which a secant cuts a curve is moved along the curve towards a second of these points, the secant approaches the position of a *tangent* to the curve at the second point, and it is by thus regarding the tangent as the limit of a secant that we are able in Analytic Geometry to obtain its equation.

A *normal* to a curve at any point is the straight line perpendicular to the tangent to the curve at that point.

As it is often necessary to consider tangents and normals to curves, we shall seek the equations of these lines for each curve that we study.

49. Let us find general forms for the equations of a tangent and a normal to a given circle at a given point of the curve.

We will use the simplest form of the equation of the circle: —

$$x^2 + y^2 = a^2, \quad \ldots\ldots\ldots\ldots \quad (1)$$

and let P_1 be the given point.

By § 26, the equation of any line through P_1 may be written

$$y - y_1 = \lambda\,(x - x_1) \quad \ldots\ldots\ldots\ldots \quad (2)$$

Equation (2) will represent a *secant*, a *tangent*, or a *normal*, according to the value given to λ.

Let λ_s represent the slope of a secant which cuts the curve in P_1 and a second point P_2; λ_t, the slope of the tangent at P_1; λ_n, the slope of the normal at P_1.

If we obtain values for λ_s, λ_t, and λ_n, in terms of known quantities, and substitute them, in turn, in equation (2), we shall have the equations of the *secant, tangent*, and *normal*, respectively.

Instead of denoting the coördinate of P_2 by (x_2, y_2) it will be convenient to introduce the following notation:

Let M_1 and M_2 be the feet of the ordinates of P_1 and P_2

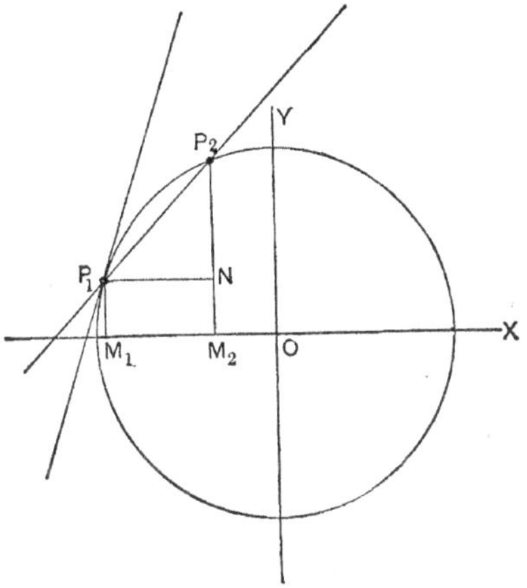

respectively, and draw through P_1 a line parallel to the axis of X cutting $M_2 P_2$ in N, and let

$$P_1 N = h, \qquad N P_2 = k.$$

According to this notation the coördinates of P_2 may be written $(x_1 + h, y_1 + k)$. The student should satisfy himself by drawing other figures that this is true for all possible positions of P_1 and P_2. The slope of the secant is

$$\lambda_s = \frac{k}{h} \quad \dots \dots \dots \dots \quad (3)$$

Now the difficulty of our problem consists in this, that as P_2 moves down the circle towards P_1 both numerator and denominator of this fraction approach zero as a limit, and it is therefore not possible to see, without further examination, what limit the fraction itself is approaching. This is not surprising, as we have not yet made any use of the fact that P_1 and P_2 are both on the circle (1). This fact gives us the two equations of condition:

$$\left.\begin{array}{r} x_1^2 + y_1^2 = a^2 \\ (x_1 + h)^2 + (y_1 + k)^2 = a^2 \end{array}\right\} \quad \ldots\ldots\ldots \quad (4)$$

By subtracting the first of these equations from the second we get the relation

$$2hx_1 + h^2 + 2ky_1 + k^2 = 0.$$

In order to get another expression for the fraction $\frac{k}{h}$ we have merely to divide this equation by h:

$$2x_1 + h + 2\frac{k}{h}y_1 + \frac{k^2}{h} = 0,$$

or

$$2x_1 + h + (2y_1 + k)\frac{k}{h} = 0.$$

Accordingly,

$$\frac{k}{h} = -\frac{2x_1 + h}{2y_1 + k}.$$

We thus have

$$\lambda_s = -\frac{2x_1 + h}{2y_1 + k} \quad \ldots\ldots\ldots \quad (5)$$

This expression for λ_s is less simple than that given by equation (3), but it has the advantage of being in a form in which we can easily determine its limit. As P_2 approaches P_1, both h and k approach zero, so that the numerator of the fraction in (5) approaches $2x_1$, the denominator $2y_1$. The slope λ_s itself therefore approaches the limit $-\frac{2x_1}{2y_1}$. Therefore

$$\lambda_t = -\frac{x_1}{y_1} \dots \dots \dots \dots (6)$$

Since the normal is perpendicular to the tangent, we have, by § 33, Cor. 2,

$$\lambda_n = -\frac{1}{\lambda_t} = \frac{y_1}{x_1} \dots \dots \dots \dots (7)$$

The equations of the required tangent and normal are, therefore,

$$y - y_1 = -\frac{x_1}{y_1}(x - x_1), \dots \dots \dots \dots (8)$$

and

$$y - y_1 = \frac{y_1}{x_1}(x - x_1). \dots \dots \dots \dots (9)$$

Clearing (8) of fractions, we have

$$y_1 y - y_1^2 = -x_1 x + x_1^2;$$

transposing and reducing by means of equation (4), we have

$$x_1 x + y_1 y = a^2, \dots \dots \dots \dots (10)$$

the simplest form of the equation of the tangent.

Reducing equation (9), we have

$$y_1 x - x_1 y = 0, \dots \dots \dots \dots (11)$$

the equation of the normal.

The form of equation (11) shows us that the normal must

pass through the origin (v. § 19). Now the origin is, in this case, the centre of the circle, so that we have proved the well-known theorem of geometry, that a perpendicular to a tangent to a circle, at the point of tangency, passes through the centre.

NOTE. This method of finding the equations of tangent and normal should be carefully studied, not only on account of the importance of the tangent and normal to the circle, but because it will be used in finding the equations of these lines for each of the curves which we shall study. Equation (1) will change as we change our curve, and so equations (4) and (5) will have different forms, and the algebraic work by which we obtain the value of λ_8 will vary; but, in the essential steps, the same method may be used to find the equations of the tangent and normal to any curve.

EXAMPLES.

(1.) Find the equations of the tangent and normal to the circle

$$x^2 + y^2 = 25$$

at the point (3, – 4).

The given point is a point of the circle, because its coördinates satisfy the equation; these coördinates therefore correspond to x_1 and y_1 in the general equations of tangent and normal, while $a^2 = 25$.

The equations sought are, therefore,

$$3x - 4y = 25,$$

and

$$-4x - 3y = 0,$$

respectively.

(2.) Find the equations of tangents and normals to the circle

$$x^2 + y^2 = 169$$

at all points of the curve which have their abscissas numerically equal to five.

(3.) Find the equation of the normal to the circle, at a point

P_1, by means of the fact that it passes through the centre; find the equation of the tangent from that of the normal by means of the mutual relation existing between these lines.

(4.) Find the equations of the tangent and normal to the circle

$$x^2 + y^2 = 2\,a\,x$$

at a point P_1 of the curve.

$$Ans. \begin{cases} (x_1 - a)\,x + y_1\,y - a\,x_1 = 0, \\ y_1\,x - (x_1 - a)\,y - a\,y_1 = 0. \end{cases}$$

(5.) Confirm these results from equations (10) and (11), by transformation of coördinates.

(6.) Obtain, by the general method, the equations of tangent and normal at P_1 to the circle

$$(x - x_c)^2 + (y - y_c)^2 = a^2.$$

$$Ans. \begin{cases} y - y_1 = -\dfrac{x_1 - x_c}{y_1 - y_c}\,(x - x_1), \\ y - y_1 = \dfrac{y_1 - y_c}{x_1 - x_c}\,(x - x_1). \end{cases}$$

(7.) Show that the same equations, obtained by transformation of coördinates from (10) and (11), are

$$(x_1 - x_c)\,(x - x_c) + (y_1 - y_c)\,(y - y_c) = a^2,$$

and

$$(y_1 - y_c)\,x - (x_1 - x_c)\,y - x_c\,y_1 + x_1\,y_c = 0.$$

(8.) Reduce the results of the last two examples to identical forms.

(9.) Find the equations of the tangent and normal to the circle

$$(x - 2)^2 + (y - 3)^2 = 10.$$

at a point (5, 4).

$$Ans. \begin{cases} 3\,x + y - 19 = 0, \\ x - 3\,y + 7 = 0. \end{cases}$$

(10.) Obtain, by the general method, the equation of the tangent at P_1 to the circle

$$x^2 + y^2 + a x + b y + c = 0.$$

$$Ans. \; x_1 x + y_1 y + \frac{a}{2}(x + x_1) + \frac{b}{2}(y + y_1) + c = 0.$$

(11.) Find the equations of the tangent and normal at the origin to the circle

$$x^2 + y^2 - 3x - 2y = 0.$$

$$Ans. \begin{cases} 3x + 2y = 0, \\ 2x - 3y = 0. \end{cases}$$

(12.) The projection on the axis of x of that part of the tangent which is contained between OX and the point of tangency is called the *sub-tangent;* in like manner the projection of that part of the normal which lies between its points of meeting with the curve and OX is called the *sub-normal.*

Find the lengths of the sub-tangent and sub-normal for the circle

$$x^2 + y^2 = a^2.$$

SUGGESTION. It follows from the definition that the sub-tangent equals the intercept of the tangent on OX, diminished by the abscissa of the point of tangency.

$$Ans. \; \frac{y_1^2}{x_1}, \; x_1.$$

(13.) Prove that the two circles given in § 45, Ex. 11, have a common tangent at the point (8, 3); show that the line which joins their centres is perpendicular to this tangent and passes through the point of tangency of the circles.

(14.) What must be the relation between a_1, b_1, and a in order that the line

$$\frac{x}{a_1} + \frac{y}{b_1} = 1$$

may touch the circle

$$x^2 + y^2 = a^2?$$

$$Ans. \; a = \frac{a_1 b_1}{\sqrt{a_1^2 + b_1^2}}.$$

(15.) When will the line

$$y = \lambda x + b$$

touch the circle

$$x^2 + y^2 = a^2?$$

Ans. When $b^2 = a^2 (1 + \lambda^2)$.

(16.) When will the line

$$x \cos \alpha + y \sin \alpha = p$$

be tangent to the circle

$$x^2 + y^2 = a^2.$$

Ans. When $p = a$.

50. The equation of the tangent to a circle from a given point without the curve may be found as follows:—

Let P_1 be the given point, and

$$x^2 + y^2 = a^2 \quad \ldots \ldots \ldots \ldots \quad (1)$$

the equation of the circle; let P' be the unknown point of tangency.

By § 49, we may write the equation of the required tangent

$$x' x + y' y = a^2, \quad \ldots \ldots \ldots \ldots \quad (2)$$

where x' and y' are at present unknown.

Since the locus of (2) must, by the conditions of the problem, contain the point P_1, the coördinates of P_1 must satisfy (2), therefore

$$x' x_1 + y' y_1 = a^2 \quad \ldots \ldots \ldots \ldots \quad (3)$$

Again, P' is by supposition a point of the circle, and, therefore, its coördinates satisfy (1), and

$$x'^2 + y'^2 = a^2 \quad \ldots \ldots \ldots \ldots \quad (4)$$

We have now two equations, (3) and (4), connecting by means of the known quantities, x_1, y_1, and a, the unknown quanties x' and y'; consequently we can find both x' and y' by eliminating between (3) and (4).

It is evident that there will be two, and only two, pairs of values for x' and y', because (3) is of the first, and (4) of the second, degree. From this we infer that two tangents can be drawn to the circle from a point without, a fact well known in Geometry. If we call the two pairs of values mentioned above (x', y') and (x'', y''), we shall have, by substituting them in turn in (2), the equations of the required tangents

$$x' \ x + y' \ y = a^2,$$
$$x'' \ x + y'' \ y = a^2,$$

touching the circle at P' and P'', respectively.

NOTE. In this discussion we have supposed P_1 a point without the circle, because we know from Geometry that no tangent can be drawn to a circle from a point within. If we apply the same method when P_1 lies within the circle, the coördinates of P' and P'' will have imaginary values, showing that there are no real tangents in this case. If P_1 lies on the circle, and the same method is used, the values of the coördinates of P' and P'' will be alike and equal to (x_1, y_1); the two tangents, therefore, will coincide in this case, a result which Geometry also teaches.

EXAMPLES.

(1.) Find the equations of the tangents from the point (7, 1) to the circle

$$x^2 + y^2 = 25.$$

Here,

$$x_1 = 7, \ y_1 = 1, \ a^2 = 25.$$

Making these substitutions in equations (3) and (4), we have

$$7 x' + y' = 25,$$

and

$$x'^2 + y'^2 = 25.$$

From these equations we find that either

$$x' = 3 \text{ and } y' = 4,$$

or

$$x' = 4 \text{ and } y' = -3.$$

The required equations are, therefore,

$$3x + 4y = 25,$$

and

$$4x - 3y = 25.$$

(2.) Find the equations of the tangents from $(-7, -1)$ to the circle

$$x^2 + y^2 = 25.$$

Ans. $\begin{cases} 3x + 4y + 25 = 0, \\ 4x - 3y + 25 = 0. \end{cases}$

(3.) What are the equations of the tangents to the same circle from the point $\left(\frac{25}{7}, \frac{25}{7}\right)$? Prove that the lengths of these tangents are equal.

(4.) Find the equation of the chord which joins the points where tangents from P_1 touch the circle

$$x^2 + y^2 = a^2.$$

This may be done by substituting the coördinates of P' and P'' in the equation of a line which passes through two points, or by the following instructive method: —

The equations of the tangents to the circle at P' and P'' are

$$x' x + y' y = a^2$$

and

$$x'' x + y'' y = a^2.$$

Since by supposition P_1 lies on each of these lines, we have

$$x' x_1 + y' y_1 = a^2$$

and

$$x'' x_1 + y'' y_1 = a^2.$$

These last equations may also be regarded as the equations of condition which make the line

$$x_1 x + y_1 y = a^2$$

pass through the points (x', y') and (x'', y''). But we are seeking the line joining P' and P''; therefore

$$x_1\, x + y_1\, y = a^2$$

is the required equation.

REMARK. If P_1 lies on the circle, this chord becomes the tangent to the circle at P_1, as is evident from its equation.

51. A *diameter* of any curve is the locus of the middle points of a set of parallel chords.

We will find the equation of a diameter of the circle

$$x^2 + y^2 = a^2 \quad . \; . \; . \; . \; . \; . \; . \; . \; . \; . \; . \; . \quad (0)$$

Let the inclination of each of the parallel chords be γ_1, and let their slope be λ_1. The equations of these chords may then be written :—

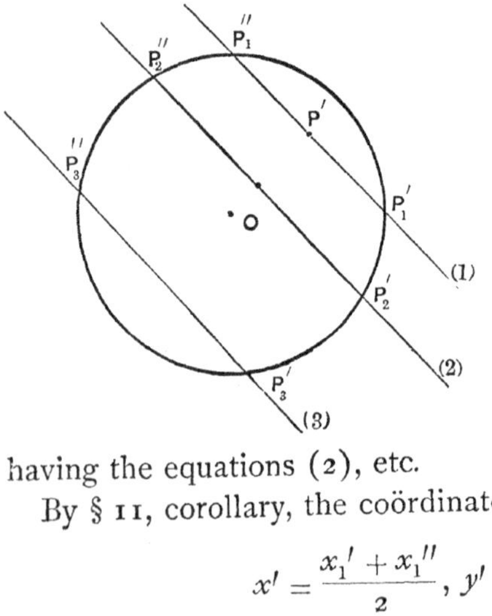

$$y = \lambda_1\, x + b_1 \quad . \; . \quad (1)$$
$$y = \lambda_1\, x + b_2 \quad . \; . \quad (2)$$
$$y = \lambda_1\, x + b_3 \quad . \; . \quad (3)$$

etc.

In the figure the line (1) represents the locus of the first of these equations; the extremities of this chord are P_1' and P_1'', and its middle point is P'. In like manner, $P_2'\, P_2''$, etc., are other chords having the equations (2), etc.

By § 11, corollary, the coördinates of P' are

$$x' = \frac{x_1' + x_1''}{2}, \quad y' = \frac{y_1' + y_1''}{2}.$$

Now P_1' and P_1'' are the points where the chord (1) cuts the circle; therefore, by § 20, their coördinates may be found by eliminating between equations (0) and (1).

Squaring (1) and substituting in (0), we have

$$x^2 + \lambda_1^2 x^2 + 2 b_1 \lambda_1 x + b_1^2 = a^2,$$

or

$$(1 + \lambda_1^2) x^2 + 2 b_1 \lambda_1 x + (b_1^2 - a^2) = 0,$$

and the two roots of this equation are x_1' and x_1''. The sum of the roots of such an equation is *minus the coefficient of x divided by the coefficient of x^2.**

Therefore

$$x' = \frac{x_1' + x_1''}{2} = -\frac{b_1 \lambda_1}{1 + \lambda_1^2}. \quad \text{. . . . . (4)}$$

Substituting this value in (1) and solving for y, we shall obtain y'; we have

$$y' = -\frac{b_1 \lambda_1^2}{1 + \lambda_1^2} + b_1;$$

reducing to a common denominator and combining we obtain

$$y' = \frac{b_1}{1 + \lambda_1^2}.$$

In the same way the coördinates of the middle point of the second chord may be found to be

$$-\frac{b_2 \lambda_1}{1 + \lambda_1^2} \quad \text{and} \quad \frac{b_2}{1 + \lambda_1^2};$$

and so for the middle point of any other chord.

Now the equation of any locus must express an ***invariable relation*** existing between the coördinates of all points on that locus. The coördinates of the different middle points only differ

* In the equation $\alpha x^2 + \beta x + \gamma = 0$ the roots are

$$\frac{-\beta \pm \sqrt{\beta^2 - 4\alpha\gamma}}{2\alpha}.$$

Accordingly the sum of the roots is $-\frac{\beta}{\alpha}$, and (a fact which we add for future reference) their product is $\frac{\gamma}{\alpha}$.

in the quantities b_1, b_2, etc., which enter as factors into the values of each coördinate; dividing y' by x' we obtain

$$\frac{y'}{x'} = -\frac{1}{\lambda_1}, \quad \ldots\ldots\ldots\ldots\ldots \quad (5)$$

a relation which is independent of b_1, and consequently must be true where P' represents the middle point of the chord (2), or (3), or any other of the system of parallel chords.

Equation (5) expresses an *invariable relation* between the coördinates of each point of the required locus, and dropping the accents we have

$$\frac{y}{x} = -\frac{1}{\lambda_1} \quad \text{or} \quad y = -\frac{1}{\lambda_1}x, \quad \ldots\ldots\ldots \quad (6)$$

the equation of the diameter which bisects all chords which have a slope λ_1.

Since equation (6) is of the *first degree*, its locus must be a straight line; as there is *no constant term*, the origin is a point of the locus.

Equation (6) may be written

$$y = \lambda x \quad \ldots\ldots\ldots\ldots\ldots\ldots \quad (7)$$

where

$$\lambda = -\frac{1}{\lambda_1}. \quad \ldots\ldots\ldots\ldots\ldots\ldots \quad (8)$$

By § 25, Remark, equation (7) represents a line through the origin with a slope λ; and, by § 33, Cor. 2, and equation (8), this line must be perpendicular to the chords which it bisects.

Note. These properties are well-known properties of the diameter of the circle, and the equation of the diameter might have been easily found by means of them; but we have used the general method of finding the diameter of any curve, in order to illustrate it by a simple example.

52. We may readily show that every chord through the centre of a circle is a diameter; for by § 25, Remark, any such chord may have its equation written in the form

$$y = \lambda x. \quad \dots\dots\dots\dots\dots \quad (1)$$

Now, whatever the value of λ, we can find a quantity λ_1, such that

$$\lambda_1 = -\frac{1}{\lambda}, \quad \text{or} \quad \lambda = -\frac{1}{\lambda_1}.$$

Since any quantity, from ∞ to $-\infty$, may be the slope of a line, it is always possible to draw a set of chords which shall have the value of λ_1, thus found, for their common slope; then, by § 51, these chords will be bisected by the locus of equation (1), and that locus must be a *diameter* of the circle.

EXAMPLES.

(1.) A system of straight lines parallel to

$$2x + 7y = 5$$

intersects the circle

$$x^2 + y^2 = 40.$$

Find the equation of the line which bisects the chords thus formed.

Here λ_1, the slope of the system of chords, is $-\frac{2}{7}$; consequently λ, the slope of the diameter, must be $\frac{7}{2}$, and the equation of the diameter is

$$y = \frac{7}{2}x,$$

or

$$7x - 2y = 0.$$

(2.) Show, by transforming coördinates, that the equation of the diameter of the circle

$$(x - x_c)^2 + (y - y_c)^2 = a^2,$$

which bisects all chords with a slope λ_1, is

$$y - y_c = -\frac{1}{\lambda_1}(x - x_c).$$

(3.) Find the equations of the diameters of the circles

$$x^2 + y^2 - 4x + 4y - 1 = 0,$$

$$x^2 + y^2 + 6x - 3y - 1 = 0,$$

which bisect all chords with an inclination of 135°.

(4.) Show that the extremities of a diameter of a circle whose centre is the origin are the points

$$\left(\frac{a}{\sqrt{1+\lambda^2}}, \frac{a\lambda}{\sqrt{1+\lambda^2}}\right), \left(-\frac{a}{\sqrt{1+\lambda^2}}, -\frac{a\lambda}{\sqrt{1+\lambda^2}}\right).$$

(5.) Prove that the lines joining the extremities of two diameters of a circle are parallel.

(6.) Prove that the tangents to a circle at opposite extremities of any diameter are parallel.

(7.) Show that the extremities of a diameter of the circle

$$(x - x_c)^2 + (y - y_c)^2 = r^2$$

are the points

$$\left(x_c \pm \frac{a}{\sqrt{1+\lambda^2}}, y_c \pm \frac{a\lambda}{\sqrt{1+\lambda^2}}\right).$$

(8.) If through one of the points of intersection of two circles a diameter of each circle is drawn, the straight line which joins the extremities of these diameters passes through the other point of intersection, and is parallel to the line joining their centres. Prove this for the circles given in § 45, Ex. 9.

53. We will close this chapter by finding the polar equation of the circle.

Let OX be the initial line, and O the pole. Let C be the centre of the circle with coördinates (r_c, ϑ_c), and let a be the radius. The coördinates of P, any point of the curve, are (r, ϑ).

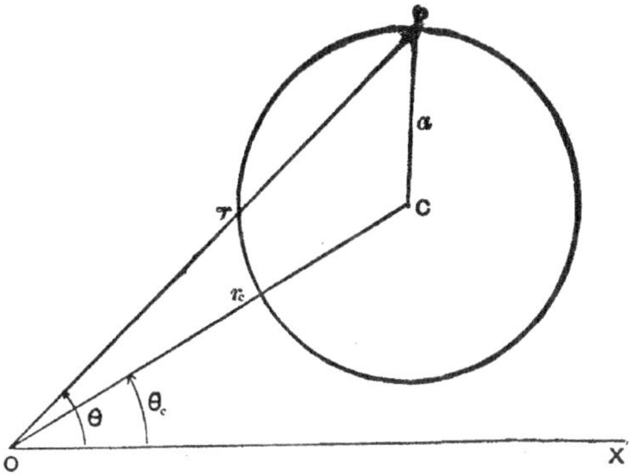

In the triangle OCP, we have by Trigonometry

$$\overline{CP}^2 = \overline{OC}^2 + \overline{OP}^2 - 2\,\overline{OC} \times \overline{OP} \times \cos COP \ldots (1)$$

Now

$$CP = a, \quad OC = r_c, \quad OP = r.$$

From the figure and § 3, we have

$$COP = XOP - XOC$$
$$= \vartheta - \vartheta_c.$$

Substituting these values in (1), we obtain

$$a^2 = r_c^2 + r^2 - 2\,r_c\,r \cos(\vartheta - \vartheta_c),$$

which must be the required equation.

This may be written in the form

$$r^2 - 2\,r_c r \cos(\vartheta - \vartheta_c) + r_c^2 - a^2 = 0.$$

It is evident from the form of this equation that with each value of ϑ there will be two values for r; a fact which agrees with the geometric properties of the circle.

Corollary 1. If C is on OX, $\vartheta_c = 0$, and the equation becomes

$$r^2 - 2\,r_c\,r \cos\vartheta + r_c^2 - a^2 = 0.$$

Corollary 2. When the pole is on the circumference, and the initial line passes through the centre,

$$r_c = a \text{ and } \vartheta_c = 0\,;$$

in this case the equation reduces to the form

$$r = 2\,a \cos \vartheta.$$

This equation gives but one value of r for each value of ; as we have divided each term by r, however, r may equal 0 wi each value of ϑ, a fact which is apparent from the hypothesis.

Corollary 3. If the centre is the pole, $r_c = 0$, and the equ tion becomes

$$r = \pm\, a,$$

as shown in § 38.

EXAMPLES.

(1.) Show that the polar equation of the circle, the orig being on the circumference and the initial line a tangent, is

$$r = 2\,a \sin \vartheta.$$

(2.) Obtain the equation of Cor. 3 from the rectangular equ tion referred to the centre, by transformation of coördinates.

(3.) Obtain the equation of Cor. 2 from that of § 45, Cor.

(4.) Obtain the general form of the polar equation from t rectangular equation

$$(x - x_c)^2 + (y - y_c)^2 = a^2.$$

(5.) Prove that the two values of r, given by the general pol equation, are real and different, real and equal, or imaginar according as $\sin(\vartheta - \vartheta_c)$ is less than $\frac{a}{r_c}$, equal to $\frac{a}{r_c}$, or great than $\frac{a}{r_c}$, in numerical value. Show that these results may al be obtained from the figure by means of well-known properti of the circle.

(6.) Prove the theorem of § 45, Ex. 13, using polar coör nates.

CHAPTER VI.

LOCI.

54. In finding the equation of the circle, we expressed in an equation its most familiar property. But the circle may also be defined by any other essential property, and we can find its equation by expressing that essential property in an equation connecting the coördinates of any point of the curve.

In § 51, a diameter of a circle is defined as the locus of the middle points of a set of parallel chords, — a definition different from the familiar one of Geometry, but expressing a well-known property of diameters. From this new definition, however, we found the invariable relation which exists between the coördinates of every point of the locus, and we obtained the equation of a diameter by expressing this invariable relation in an equation. The result was recognized as representing a straight line through the centre, and in this way we obtained from the equation the properties by which a diameter is commonly defined. If these properties were unknown to us, we should be no less certain of their truth.

In like manner, if a point moves according to a given geometric law, we can find the equation of its path, or locus, though ignorant of the form of that path. This equation may be recognized as the equation of some known curve, in which case we can determine at once the locus of the moving point; if, however, the equation is not familiar, we can deduce from it the form and properties of the curve.

In this chapter we shall apply this method to several loci defined by properties more complicated than those mentioned above, and we shall add a number of other examples leaving the work wholly or in part to the student.

55. Let us find the locus of the vertex of a triangle when the base and the difference of the squares of the sides are given.

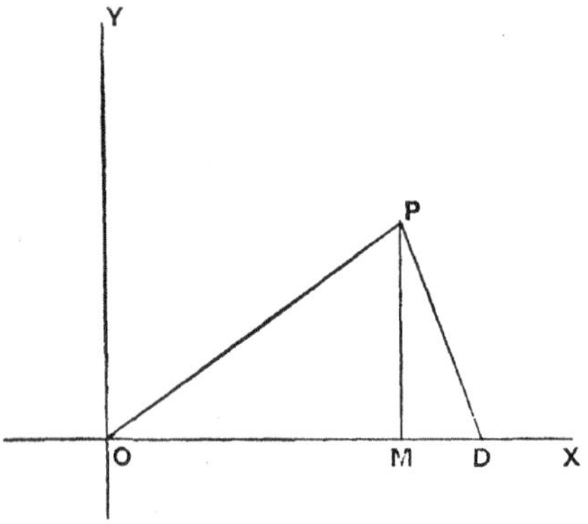

Take the base for the axis of x, and its left-hand extremity for the origin. Call the vertex P, with coördinates, $x = OM$, $y = MP$; let the length of the base, OD, be c, and let m^2 represent the constant difference between the squares of the sides.

The definition of the locus may be expressed by the equation

$$\overline{OP}^2 - \overline{DP}^2 = m^2, \quad \ldots\ldots\ldots\ldots (1)$$

which will be the equation of the required locus when we express OP and DP in terms of x, y, and c.

Since the coördinates of D are $(c, 0)$, we may write, by § 10,

$$\overline{OP}^2 = x^2 + y^2,$$

$$\overline{DP}^2 = (c - x)^2 + y^2.$$

Substituting these values in equation (1), we have

$$x^2 + y^2 - [(c - x)^2 + y^2] = m^2,$$

which may be reduced to the form

$$2cx - c^2 = m^2,$$

or

$$x = \frac{c^2 + m^2}{2c}.$$

From the form of this equation, we know that the locus is a straight line parallel to OY, and therefore perpendicular to the base of the triangle, at a distance from O equal to $\frac{c^2 + m^2}{2c}$.

It should be noticed that the other segment of the base is $\frac{c^2 - m^2}{2c}$, a result which may also be obtained by taking the origin at D.

(2.) Show that if the sum of the squares of the sides is given instead of their difference, the locus is a circle with the centre at the middle of the base. Find the length of the radius of this circle.

(3.) Solve problems (1) and (2), taking the origin at the middle of the base (v. Ex. 5, Fig.), and show that the results agree with those already found.

(4.) Given the base of an isosceles triangle, find the locus of the vertex.

Ans. A straight line perpendicular to the base and bisecting it.

(5.) Given the base and the ratio of the sides of a triangle, find the locus of the vertex.

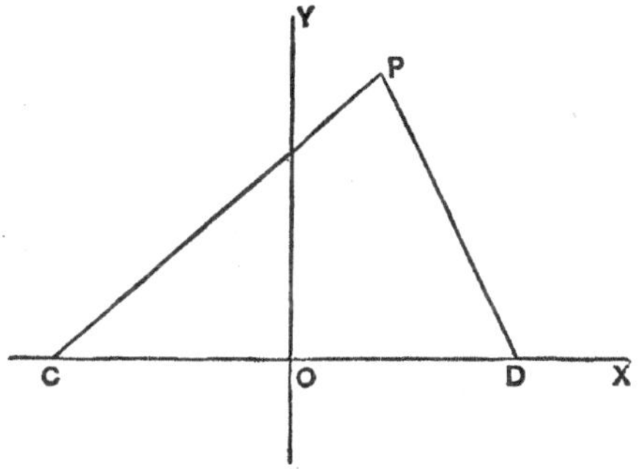

Let the base be the axis of x, and its middle point the origin. Let the length of the base equal $2c$, and the ratio of CP to DP be $m : n$.

With this notation, the coördinates C and D are $(-c, 0)$ and $(c, 0)$; the vertex P has coördinates (x, y). We have, by § 10,

$$CP = \sqrt{(x + c)^2 + y^2},$$
$$DP = \sqrt{(x - c)^2 + y^2}.$$

The equation of the required locus may, therefore, be writte[n]

$$\frac{\sqrt{(x+c)^2+y^2}}{\sqrt{(x-c)^2+y^2}}=\frac{m}{n};$$

squaring and clearing of fractions, this becomes

$$m^2\,[(x-c)^2+y^2]=n^2\,[(x+c)^2+y^2],$$

or

$$(m^2-n^2)\,x^2-2\,(m^2+n^2)\,c\,x+(m^2-n^2)\,y^2+(m^2-n^2)\,c^2=c$$

which may be written

$$x^2-2\,\frac{m^2+n^2}{m^2-n^2}\,c\,x+y^2+c^2=0.$$

This may be readily shown to be the equation of a circle, wit[h] its centre at the point $\left(\frac{m^2+n^2}{m^2-n^2}\,c,\ 0\right)$ and its radius equal t[o] $\frac{2\,m\,n}{m^2-n^2}\,c.$

(6.) Given the base of a triangle, and m times the square c[of] one side plus n times the square of the other; find the locus c[of] the vertex.

Ans. Using the same axes and notation as in the last exampl[e] the locus is a circle with its centre at $\left(\frac{m-n}{m+n}\,c,\ 0\right)$.

(7.) Given the base and sum of the sides of a triangle; fin[d] the equation of the locus of the vertex.

Ans. If $2\,c$ is the base and $2\,a$ the sum of the varying side[s] and the axes as in Ex. 5, we have for the equation of the locus

$$\sqrt{(x+c)^2+y^2}+\sqrt{(x-c)^2+y^2}=2\,a.$$

Removing one radical to the second member, and squarin[g] twice, we shall have for the required equation

$$(a^2-c^2)\,x^2+a^2\,y^2=a^2\,(a^2-c^2).$$

This is neither the equation of a straight line nor of a circle[;] we shall, however, have occasion to study its locus hereafter.

(8.) Show that if the *difference* of the sides in the last exam-le is given equal to $2a$, instead of their sum, the equation of he locus has the same form.

(9.) Find the locus of a point such that, if straight lines be rawn to it from the four corners of a given square, the sum of heir squares is constant.

Ans. A circle whose centre is the centre of the square, and hose radius is $\sqrt{2(m^2 - a^2)}$, where each side of the square is a and the given constant is represented by $8m^2$.

(10.) Find the locus of a point whose distances from two traight lines, given by their equations, have a given ratio, $m : n$.

Let the equations of the lines be

$$Ax + By + C = 0, \quad \ldots\ldots\ldots\ldots \quad (1)$$
$$A'x + B'y + C' = 0; \quad \ldots\ldots\ldots\ldots \quad (2)$$

nd let P' with coördinates (x', y') be the moving point.

The distances of P' from lines (1) and (2) are, by § 36,

$$\frac{Ax' + By' + C}{\sqrt{A^2 + B^2}} \quad \text{and} \quad \frac{A'x' + B'y' + C'}{\sqrt{A'^2 + B'^2}}.$$

Putting m times one of these equal to n times the other, we ave the equation of the locus, where x' and y' are the variable oördinates. Clearing of fractions and writing x for x' and y or y', we evidently have an equation of the *first degree;* there-ore the required locus is a straight line.

(11.) Show that the equations of the bisectors of the angles etween the lines

$$Ax + By + C = 0 \quad \text{and} \quad A'x + B'y + C' = 0$$

re

$$\frac{Ax + By + C}{\sqrt{A^2 + B^2}} = \pm \frac{A'x + B'y + C'}{\sqrt{A'^2 + B'^2}}.$$

(12.) A point moves so that the sum of the squares of its istances from the four sides of a given square is constant;

show that the locus of the point is a circle; find its centre and radius.

(13.) Find the locus of a point, the square of whose distance from a given point is proportional to its distance from a given line.

Ans. A circle.

SUGGESTION. Take the fixed line for the axis of y, and the perpendicular upon it from the fixed point for the axis of x.

(14.) Find the equation of the locus of a point whose distance from a fixed point always equals its distance from a fixed line.

Ans. $y^2 = 2\,x_1\,x - x_1^2$ [axes as in 13].

56. There are many problems in which it is difficult to express at once the given conditions in terms of the coördinates of the moving point and known quantities. In such cases we may introduce other variable quantities, and then find enough equations expressing necessary relations of the parts of the figure to enable us to eliminate the variable quantities thus introduced. We shall then have an equation which fulfils the requirements of the definition of the equation of a curve.

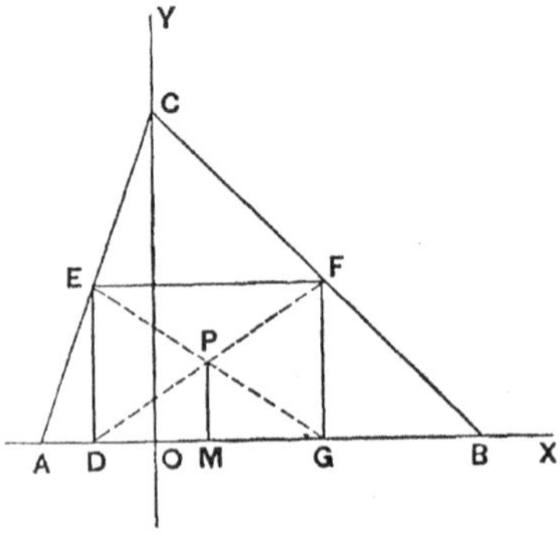

(1) To illustrate, let us find the locus of the middle points of rectangles inscribed in a given triangle.

We will take the base and perpendicular from the vertex as axes, and will suppose the triangle given by means of the intercepts which the sides cut from these axes.

Let

$$OA = a, \quad OB = a', \quad OC = b.$$

Let $D\,E\,F\,G$ represent any rectangle inscribed in the triangle, and let its middle point be P.

We can easily express the coördinates of P in terms of those of E and F; for, from the figure and § 11, it is evident that x must equal the half sum of x_E and x_F (the abscissas of E and F), and y, one-half the common ordinate of these points. Call this common ordinate k; then the equation of $E\,F$ may be written

$$y = k. \qquad \qquad (1)$$

By § 22, the equations of $A\,C$ and $B\,C$ are

$$\frac{x}{a} + \frac{y}{b} = 1, \qquad \qquad (2)$$

and

$$\frac{x}{a'} + \frac{y}{b} = 1 \qquad \qquad (3)$$

Since E and F are the points where the line (1) meets the lines (2) and (3), we have by § 20

$$x_E = \frac{a\,(b - k)}{b}, \qquad x_F = \frac{a'\,(b - k)}{b};$$

therefore

$$x = \frac{x_E + x_F}{2} = \frac{(a + a')\,(b - k)}{2\,b}; \qquad \qquad (4)$$

also for the point P

$$y = \frac{k}{2}. \qquad \qquad (5)$$

Equations (4) and (5) contain the coördinates of P, the variable k, and the known constants a, a' and b. We may, therefore, by combining these equations eliminate k and obtain an equation connecting x and y by means of the given quantities a, a', and b. This equation is

$$x = \frac{(a + a')\,(b - 2\,y)}{2\,b},$$

or

$$\frac{2x}{a+a'}+\frac{2y}{b}=1,$$

the equation of a straight line bisecting the base and perpendicular of the triangle, for its intercepts are

$$\frac{a+a'}{2} \text{ and } \frac{b}{2}.$$

REMARK. The student will remember that, in finding the equation of a diameter of a circle, we first expressed the coördinates of any point of the locus in terms of another variable, the intercept of any chord on OY. That problem, therefore, is one of the class we are now considering.

(2.) A line is drawn parallel to the base of a given triangle and its extremities joined transversely to those of the base; find the locus of the point of intersection of the joining lines.

With the same axes and notation as in the last example, the equations of the joining lines may be formed by § 27. Finding the coördinates of the point where these lines meet, we have, after eliminating k, the equation of the locus,

$$\frac{2x}{a+a'}+\frac{y}{b}=1,$$

which represents the medial line drawn from the vertex of the triangle.

(3.) A line of constant length moves with its extremities on two straight lines at right angles to each other; find the locus of its middle point.

This may be done conveniently by introducing as a third variable the angle which the moving line makes with one of the fixed lines.

Ans. A circle. Compare § 11, Ex. 9.

(4.) A square is moved so as always to have the two extremities of one of its diagonals upon two fixed straight lines at right angles to each other; show that the extremities of the other diagonal will at the same time move upon two other fixed straight lines at right angles to each other.

(5.) From one extremity A' of a fixed diameter $A'A$ of a given circle, a secant is drawn through any point P' of the circumference; at P' a tangent is drawn to the circle, and a perpendicular to the tangent is drawn from A, and extended to meet the secant in P. Find the locus of P.

Take the fixed diameter as the axis of x, and the centre as origin. We shall use as auxiliary variables the coördinates of P'. The points A' and A have coördinates $(-a, 0)$ and $(a, 0)$.

We can now write the equation of $A'P'$ by § 27, and of AP by § 33, Cor. 2; for we know the slope of the tangent at P' by § 49. Since P is the point of intersection of $A'P$ and AP, its coördinates may be found by § 20, and will be

$$x = 2x' + a, \quad y = 2y',$$

Expressing x' and y' in terms of x, y, and a, by means of these equations, and substituting the values thus found in the equation of condition which puts P' upon the circle,

$$x'^2 + y'^2 = a^2,$$

we have for the equation of the locus

$$(x - a)^2 + y^2 = 4a^2,$$

which represents a circle with its centre at A, and AA' for its radius.

(6.) Given the base of a triangle and the length of the medial line drawn from one of its extremities; find the locus of the vertex.

Take the base as the axis of x, and the end from which the medial line is drawn as the origin. Introduce as auxiliary variables the coördinates of the other extremity P' of the given medial line.

Ans. A circle with centre at the point $(-c, 0)$ and radius $2m$, where c and m represent the lengths of the base and medial line.

(7.) To find the locus of the centre of a circle which has a given radius and passes through a given point.

Let P' be the unknown centre. The equation of the circle may then be written

$$(x - x')^2 + (y - y')^2 = a^2.$$

Making this pass through the given point P_1, we have

$$(x_1 - x')^2 + (y_1 - y')^2 = a^2;$$

changing the signs of the quantities in the parentheses (but not of their squares) and writing x and y for the variables x' and y', this equation becomes

$$(x - x_1)^2 + (y - y_1)^2 = a^2,$$

the equation of a circle with P_1 for the centre, and a for its radius.

REMARK. This equation may be easily obtained from the fact that the distance of the moving centre from the fixed point must be equal to the given radius; but the method used above is instructive.

(8.) Find the locus of the middle points of chords of a given circle, drawn from a fixed point on the circumference.

Ans. A circle with the radius of the given circle for its diameter.

Compare with § 45, Ex. 13.

57. Instead of expressing in an equation the geometrical definition of the locus, and then substituting for each quantity in this equation its value in terms of the coördinates of the moving point and known quantities, it is often more convenient to use as the primary equation one which expresses some necessary relation of the parts of the figure other than the one explicitly

given. From our hypothesis we may be able to express all quantities in this equation in terms of the variable coördinates and known quantities, and so find the equation of the locus.

We will illustrate by an example in which this method is used in connection with that of § 56.

(1.) Given the base and sum of the sides of a triangle, if the perpendicular be produced beyond the vertex until its whole length is equal to one of the sides, find the locus of the extremity of the perpendicular.

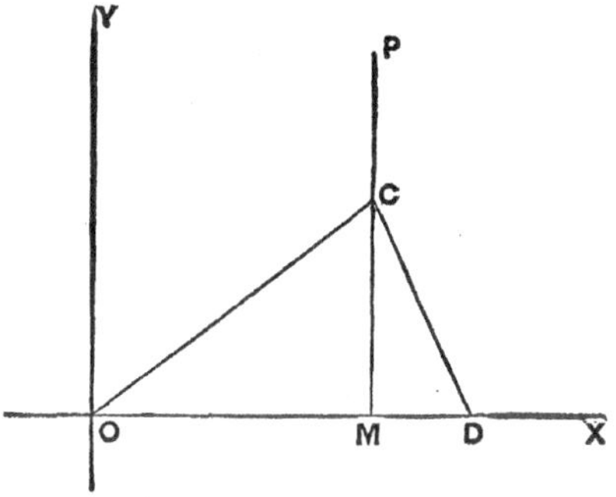

Take the axes as in § 55, Ex. 1; let the base be c, and the sum of the varying sides be m; let MP be made equal to OC, whatever the position of C.

We may, therefore, write

$$OC = MP = y,\quad CD = m - y.$$

Now, from the figure, it is evident that

$$\overline{MC}^2 = \overline{OC}^2 - \overline{OM}^2 = y^2 - x^2,$$

and

$$\overline{MC}^2 = \overline{CD}^2 - \overline{MD}^2 = (m - y)^2 - (c - x)^2.$$

Putting these values of $\overline{MC}^2$ equal to each other, we have the expression of a necessary relation between x and y, in an equation which contains only those quantities and the given constants m and c. Expanding and reducing, this equation becomes

$$2cx - 2my + m^2 - c^2 = 0,$$

which is of the first degree, and therefore represents a straight line.

(2.) A line of constant length moves with its ends on two straight lines at right angles to each other, and perpendiculars to the lines are raised at its extremities; find the locus of their intersection.

Ans. A circle.

58. Polar coördinates may often be employed to advantage.

(1.) To find the locus of the middle points of chords of a circle, which pass through a given point.

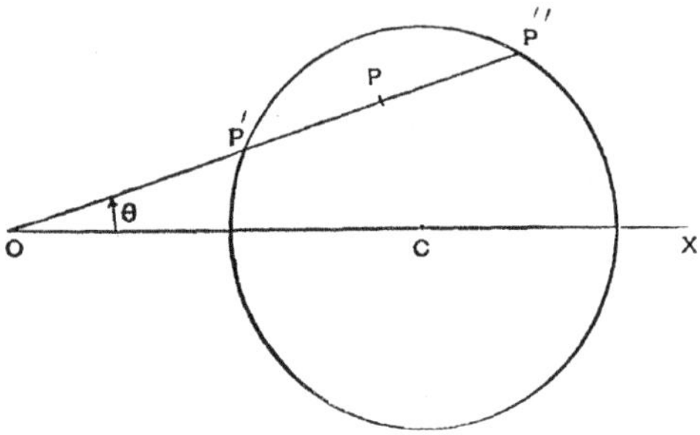

Let the given point be the pole, and the line joining it with the centre be the initial line. Let $P' P''$ be any chord which passes through O, and let P be its middle point, with coördinates r and ϑ.

It is evident from the figure that ϑ will be the same for P, P', and P'', and that

$$r = \frac{r' + r''}{2} \quad \ldots\ldots\ldots\ldots \quad (1)$$

The equation of the circle may be written, by § 53, Cor. 1,

$$r^2 - 2\, r_c\, r \cos \vartheta + r_c^2 - a^2 = 0. \quad \ldots\ldots\ldots \quad (2)$$

When ϑ equals $X O P''$, the roots of this equation are r' and r''; therefore, by equation (1),*

$$r = r_c \cos \vartheta, \quad \ldots\ldots\ldots\ldots \quad (3)$$

and similarly with every value of ϑ; for though for certain values of ϑ, r' and r'' are imaginary, yet r, their half-sum, is always real and equal to $r_c \cos \vartheta$.

Equation (3) is, therefore, the equation of the required locus, which, by § 53, Cor. 2, must be a circle with r_c for a diameter.

* V. foot-note on p. 90.

(2.) Find the locus of the middle points of chords drawn from the extremity of any diameter of a circle.

Compare with § 56, Ex. 8.

(3.) Find the locus of a point which moves so as to divide in a given ratio all lines joining a fixed point with a fixed straight line.

59. A great many problems in loci lead us to equations of an unfamiliar form. We have had an illustration of this in Ex. 7, on page 99. In such a case the locus will usually, as in the example just referred to, not come under any of the forms of curves we have so far studied. There is, however, one method by which we can sometimes determine what the locus of an unfamiliar equation is.

Consider, for example, the equation

$$xy = 0 \quad \ldots \ldots \ldots \ldots \quad (1)$$

which is in an unfamiliar form since it is of the second degree and yet is not of the form of the equation of a circle (§ 46). It is clear that any point whose abscissa or whose ordinate is zero lies on the curve (1), but that no other points lie on it. The locus of (1) therefore consists of two parts: first the axis of X, second the axis of Y.

As a second example let us consider the equation

$$x^2 - xy = 0 \quad \ldots \ldots \ldots \ldots \quad (2)$$

This equation may be written

$$x(x - y) = 0 \quad \ldots \ldots \ldots \ldots \quad (3)$$

It will therefore be satisfied by the coördinates of any point whose abscissa is zero, i.e., of any point on the axis of Y. There is, however, another way in which (3) can be satisfied, namely by the equality of x and y. The locus of (2) or (3) therefore consists of two parts: first the axis of Y, $x = 0$; second the straight line $x - y = 0$. Moreover it is clear that a point not lying on either of these two lines does not belong to the locus, since for such a point neither x nor $x - y$ is zero, and therefore (3) is not satisfied.

The general principle of which the two foregoing examples are illustrations may be stated as follows: Suppose that all the terms of the equation we wish to investigate have been transposed to the left-hand side, and suppose that this left-hand side can then be factored into two factors which we will denote by the letters u and v. The equation we wish to investigate may therefore be written

$$uv = 0 \quad . \quad . \quad . \quad . \quad . \quad . \quad . \quad . \quad . \quad . \quad . \quad . \quad . \quad (4)$$

where we must not forget that the letters u and v stand for certain more or less complicated expressions involving the variables x and y. *Then the curve* (4) *consists of two and only two parts, first the curve*

$$u = 0 \quad . \quad . \quad . \quad . \quad . \quad . \quad . \quad . \quad . \quad . \quad . \quad . \quad . \quad . \quad (5)$$

and second the curve

$$v = 0 \quad . \quad . \quad . \quad . \quad . \quad . \quad . \quad . \quad . \quad . \quad . \quad . \quad . \quad . \quad (6)$$

Let (x_1, y_1) be any point on the curve (5); (x_2, y_2) any point on the curve (6); and (x_3, y_3) any point which does not lie on either of the curves (5) or (6). The statement we have made will be proved if we can show that (x_1, y_1) and (x_2, y_2) both lie on (4), but (x_3, y_3) does not.

In order to find out whether (x_1, y_1) lies on (4) or not, we must replace the variables (x, y) in (4) by the constant values (x_1, y_1) and then see whether the equation is fulfilled. In doing this we must remember that u is merely an abbreviation for a certain expression in the variables x and y. This expression will therefore take on a certain constant value, which we will call u_1, when these variables are replaced by the constants x_1, y_1. In the same way the expression v takes on a constant value v_1 when x, y are replaced by x_1, y_1. Thus equation (4) takes on, after this substitution, the form

$$u_1 v_1 = 0 \quad . \quad . \quad . \quad . \quad . \quad . \quad . \quad . \quad . \quad . \quad . \quad . \quad . \quad (7)$$

and it remains to be proved that this is a true equation. By hypothesis (x_1, y_1) lies on the curve (5). Accordingly its coördinates satisfy (5) and we have

$$u_1 = 0,$$

and from this the truth of (7) follows.

We have thus proved that (x_1, y_1) lies on (4), and by precisely the same method we show that (x_2, y_2) does so.

In order finally to show that (x_3, y_3) does not lie on (4) let us indicate by u_3 the value of the expression u when the variables x, y are replaced by the constants x_3, y_3, and by v_3 the value of v after the same substitution. The result of substituting x_3, y_3 in (4) is then

$$u_3 v_3 = 0 \quad . \quad . \quad . \quad . \quad . \quad . \quad . \quad . \quad . \quad . \quad . \quad . \quad . \quad (8)$$

This equation, however, is not true, since, the point (x_3, y_3) lying on neither (5) nor (6), neither u_3 nor v_3 is zero.

The result of this section may be stated by saying that if we have two curves whose equations are so written that the right-hand members are zero,* the two curves may be represented by a single equation obtained by multiplying these equations together.

* This restriction is very important, for if the right-hand members are not both zero the fact here stated will not be true. Thus if we have the two lines $x = 1$, $x = 2$ and multiply these equations together as they stand, the resulting equation $x^2 = 2$ represents the two wholly different lines $x = \pm\sqrt{2}$. In order to get an equation representing the two lines we started with we must write their equations $x - 1 = 0$, $x - 2 = 0$, and then multiply; getting $x^2 - 3x + 2 = 0$ as the equation of the pair of lines.

EXAMPLES.

What are the loci of the following equations ?—

(1.) $x^2 - y^2 = 0$. (2.) $x^2 + 2xy - 3x = 0$.

(3.) $x^2y + y^3 - 4y = 0$. (4.) $x^4 - y^4 - 9x^2 + 9y^2 = 0$.

(5.) Prove that if u, v, w are three expressions in x, y, the equation $uvw = 0$ has as its locus the three curves: $u = 0$, $v = 0$, $w = 0$.

(6.) Show that the first members of the following equations can be factored:

$$ax^2 + bx + c = 0,$$

$$ax^2 + bxy + cy^2 = 0.$$

What do these equations represent ?

60. In the last section we have seen what the result is of multiplying together the two equations

$$u = 0 \quad . \quad . \quad . \quad . \quad . \quad . \quad . \quad . \quad . \quad . \quad . \quad . \quad . \quad (1)$$

$$v = 0 \quad . \quad . \quad . \quad . \quad . \quad . \quad . \quad . \quad . \quad . \quad . \quad . \quad . \quad (2)$$

Let us now consider what the result of adding these equations is. That is, we wish to find out as much as we can about the locus of the equation

$$u + v = 0 \quad . \quad . \quad . \quad . \quad . \quad . \quad . \quad . \quad . \quad . \quad (3)$$

Let (x_1, y_1) be any point common to the two curves (1) and (2), i.e. any point of intersection or of contact of these curves, and let us, as in the last section, denote by u_1 and v_1 the values of the expressions u and v when the variables x, y are replaced by the constant values x_1, y_1. Now since (x_1, y_1) lies on (1) we have

$$u_1 = 0.$$

Since (x_1, y_1) lies on (2) we have

$$v_1 = 0.$$

Accordingly

$$u_1 + v_1 = 0,$$

i.e. (x_1, y_1) lies on (3). Thus we see that (3) passes through all the points common to (1) and (2).

Now let (x_2, y_2) be a point of (1) which does not lie on (2). Then, letting u_2, v_2 denote the values of u and v when the variables x, y are given the values x_2, y_2, we have

$$u_2 = 0,$$

$$v_2 \neq 0.$$

Therefore

$$u_2 + v_2 \neq 0.$$

That is (x_2, y_2) does not lie on (3).

In the same way we see that a point (x_3, y_3) which lies on (2) but not on (1) will not lie on (3).

Thus we have established the fact that if we add together the equations of two curves, the resulting equation will have a locus which passes through all the points common to the two curves, but meets neither curve in any other point.

We can easily throw this result into a slightly more general form. The equations of the curves (1) and (2) may, if we wish, be written in the form

$$a\,u = 0 \quad \ldots\ldots\ldots\ldots\ldots \quad (4)$$

$$b\,v = 0 \quad \ldots\ldots\ldots\ldots\ldots \quad (5)$$

where a, b are any constants other than zero. By applying the principle just established to the curves (4) and (5) we get the theorem :

If $u = 0$ *and* $v = 0$ *are the equations of two given curves, then*

$$a\,u + b\,v = 0$$

represents a curve which passes through all the points common to the two given curves, and meets neither of them in any other point.

Just what curve of this sort we get depends, of course, on the values of the constants a and b.

61. As an illustration of the way in which this principle can be applied let us consider the problem of finding the equation of the common chord of two intersecting circles. Consider, for instance, the two circles.

$$x^2 + y^2 - 2x - 4y + 4 = 0 \quad \ldots\ldots\ldots \quad (1)$$

$$x^2 + y^2 - 4x - 2y + 1 = 0 \quad \ldots\ldots\ldots \quad (2)$$

These circles intersect, as the reader will see if he draws a rough figure. In order to find the equation of their common chord let us denote the first member of (1) by u, the first member of (2) by v. Then the equation

$$a\,u + b\,v = 0 \quad \ldots\ldots\ldots\ldots \quad (3)$$

will represent some curve through the two points of intersection of (1) and (2). But it is clear that, in general, the equation (3) is of the second degree and therefore does not represent a straight line. If, however, we give to a the value 1, to b the value -1, the terms of the second degree cancel and the equation (3) becomes

$$u - v = 0$$

or

$$2x - 2y + 3 = 0 \quad \ldots\ldots\ldots\ldots \quad (4)$$

This, being of the first degree, represents a straight line, and, being merely a special case of (3), this line must pass through the points of intersection of (1) and (2). Thus (4) is the equation we were seeking.

The equation (4) was obtained by subtracting the equation of one of the given circles from the other; and it is clear that the same reasoning applies to any similar case. That is, to find the equation of the common chord of two given intersecting circles, write the equations of these circles so that the coefficients of x^2 and y^2 are 1, and then subtract one equation from the other.

EXAMPLES.

(1.) A circle has its centre at the origin and radius 1, and a second circle has its centre at the point $(-1, 2)$ and radius 2. Find the equation of the common chord of these circles.

(2.) Show that the three lines

$$2x + y - 3 = 0,$$

$$x - 2y + 5 = 0,$$

$$4x - 3y + 7 = 0$$

meet in a point.

[SUGGESTION: Show that if the first two equations are denoted by $u = 0$, $v = 0$, the third can be written in the form $au + bv = 0$.]

(3.) Find the equation of the line which connects the origin with the point of intersection of the lines

$$x - 2y - 3 = 0,$$

$$2x - y + 5 = 0.$$

[SUGGESTION: Determine the constants a and b in the equation $au + bv = 0$, so as to eliminate the constant term.]

(4.) Prove that the three common chords of three intersecting circles meet in a point.

[SUGGESTION: Let the circles be

$$x^2 + y^2 + a_1 x + b_1 y + c_1 = 0,$$

$$x^2 + y^2 + a_2 x + b_2 y + c_2 = 0,$$

$$x^2 + y^2 + a_3 x + b_3 y + c_3 = 0.$$

Find their common chords by § 61, and then use the method of example 2.]

CHAPTER VII.

THE CONIC SECTIONS.

62. *An ellipse* is a curve generated by a point moving in a plane so that *the sum of its distances from two fixed points in that plane is constant.*

An hyperbola is a curve generated by a point moving in a plane so that the *difference of its distances from two fixed points in that plane is constant.*

A parabola is a curve generated by a point moving in a plane so that at each instant *its distances from a fixed point and a fixed line in that plane are equal.*

These three curves are called *the conic sections*, because, if a right circular cone is cut by a plane which does not pass through the vertex of the cone, the section will be bounded by one of these curves.

We will find the equations of the conic sections.

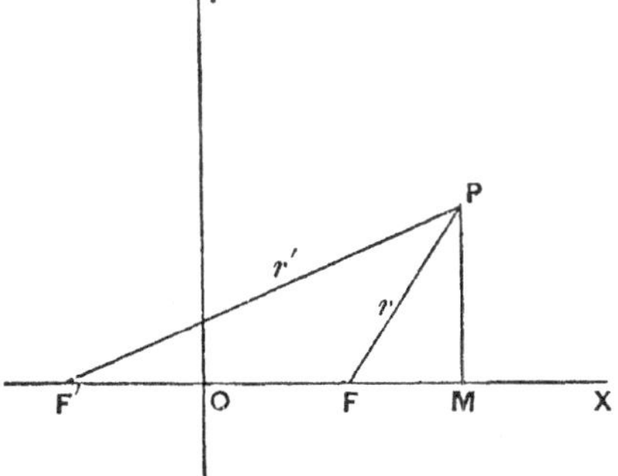

63. In finding the equation of the *ellipse*, we will take the straight line joining the two fixed points, which are called *foci*, for the axis of x, and the point midway between the foci, called the *centre*, for the origin.

In the figure, F' and F are the foci; P is any point

of the curve; and r' and r, called *focal radii*, measure the distances of P from F' and F, respectively.

Let $2c$ represent the constant distance between the foci, and $2a$ the sum of the focal radii, which is constant by the definition of the curve. In the triangle $F'FP$ the side $F'F = 2c$ is less than $2a$, the sum of the other two sides. Therefore

$$c < a \quad \ldots\ldots\ldots\ldots\ldots \quad (1)$$

The coördinates of F' and F are $(-c, 0)$ and $(c, 0)$, and the equation which expresses the definition of the locus is

$$r' + r = 2a.$$

Expressing r' and r in terms of x, y, and c, by § 10,

$$r' = \sqrt{(x+c)^2 + y^2}, \quad \ldots\ldots\ldots \quad (2)$$

$$r = \sqrt{(x-c)^2 + y^2} \quad \ldots\ldots\ldots \quad (3)$$

Substituting these values, we have the equation of the ellipse,

$$\sqrt{(x+c)^2 + y^2} + \sqrt{(x-c)^2 + y^2} = 2a, \quad \ldots\ldots \quad (4)$$

which will be in a more convenient form when cleared of radicals.

Equation (4) may be written

$$\sqrt{(x+c)^2 + y^2} = 2a - \sqrt{(x-c)^2 + y^2};$$

squaring,

$$(x+c)^2 + y^2 = 4a^2 - 4a\sqrt{(x-c)^2 + y^2} + (x-c)^2 + y^2;$$

expanding, combining, and transposing,

$$4a\sqrt{(x-c)^2 + y^2} = 4a^2 - 4cx;$$

dividing by 4, and squaring again,

$$a^2x^2 - 2a^2cx + a^2c^2 + a^2y^2 = a^4 - 2a^2cx + c^2x^2;$$

reducing,

$$(a^2 - c^2)x^2 + a^2y^2 = a^2(a^2 - c^2). \quad \ldots\ldots\ldots \quad (5)$$

We will represent the constant quantity $(a^2 - c^2)$, which, as we see from (1), is positive, by b^2; with this notation, (5) becomes

$$b^2 x^2 + a^2 y^2 = a^2 b^2, \quad \ldots\ldots\ldots\ldots (6)$$

', dividing by $a^2 b^2$,

$$\frac{x^2}{a^2} + \frac{y^2}{b^2} = 1, \quad \ldots\ldots\ldots\ldots (7)$$

e usual form of the equation.

64. With the same choice of axes, the *equation of the hyper-*
la may be found in a similar manner.

In this case we represent the constant *difference* between the
ocal radii by $2a$. Since the side $F'F = 2c$ of the triangle
'FP is greater than the difference $2a$ of the other two
des, we have

$$c > a \quad \ldots\ldots\ldots\ldots (1)$$

ccordingly we use b^2 as an abbreviation for $c^2 - a^2$.

With this notation show that the equation of the hyperbola is

$$\frac{x^2}{a^2} - \frac{y^2}{b^2} = 1. \quad \ldots\ldots\ldots\ldots (2)$$

NOTE. Since the equations of the ellipse and hyperbola differ
ly in the sign of b^2, the results of similar operations performed
each will differ in the same way. One of these results may,
erefore, be obtained from the other by changing the sign of b^2,
method which will enable us to infer many properties of the
perbola from the corresponding properties of the ellipse, or
ce versa.

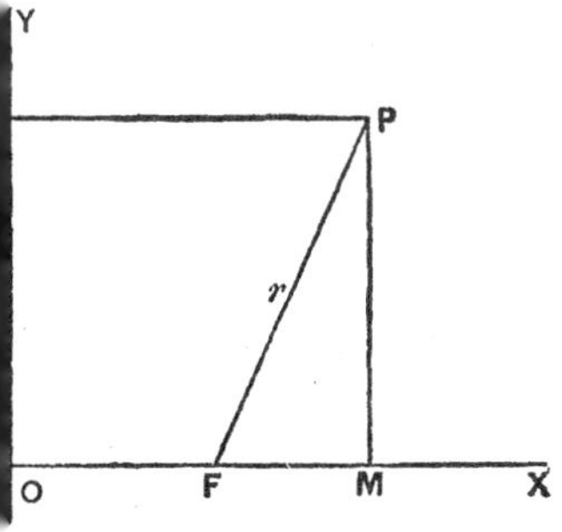

65. In finding the equation of the *parabola*, we will take the fixed straight line, called the *directrix*, as the axis of y, and the perpendicular upon it from the fixed point, or *focus*, as the axis of x.

In the figure OY is the directrix, F the focus, and P any point of the curve. Letting m

represent the constant distance of the focus *from* the directrix, the coördinates of F are $(m, 0)$.

The equation which expresses the definition of the locus is

$$FP = EP. \quad \ldots\ldots \quad (1)$$

By § 10

$$FP = \sqrt{(x-m)^2 + y^2},$$

Also

$$EP = OM = x.$$

Substituting these values in (1), we have, after squaring,

$$(x-m)^2 + y^2 = x^2, \quad \ldots\ldots \quad (2)$$

which reduces to

$$y^2 - 2mx + m^2 = 0, \quad \ldots\ldots \quad (3)$$

the equation sought.

66. A simpler form of the equation of the parabola may be found by changing the origin to the point midway between the focus and directrix.

The formulas for transformation are

$$x = x' + \frac{m}{2}, \quad y = y',$$

and the new equation is, therefore,

$$y'^2 - 2m\left(x' + \frac{m}{2}\right) + m^2 = 0,$$

which becomes, after reducing and dropping the accents,

$$y^2 = 2mx,$$

the usual form of the equation of the parabola.

NOTE. The student will remember that we have already found the equations of the conic sections in the chapter on loci. It will be useful for him to compare the definitions of the conics and their equations with examples 7, 8, 14, of § 55.

67. We will now investigate the forms of the conic sections by discussing their equations.

Beginning with the ellipse, we find its intercepts on OX to be a and $-a$; on OY, b and $-b$. Solving equation (6), § 63, for y, we have

$$y^2 = \frac{b^2}{a^2}(a^2 - x^2),$$

or

$$y = \pm \frac{b}{a}\sqrt{a^2 - x^2}, \quad \ldots\ldots\ldots\ldots (1)$$

from which it appears that with each value of x there are two values of y, *equal numerically* but *opposite in sign;* this shows that the curve is symmetrical with respect to OX. Similarly solving for x, we have

$$x = \pm \frac{a}{b}\sqrt{b^2 - y^2} \quad \ldots\ldots\ldots\ldots (2)$$

From this equation we infer that the curve is symmetrical with respect to OY.

From (1) we see that the values of y are real when x is less than a, null when x equals a, and imaginary when x is greater than a, in numerical value; also, as x increases, y must decrease.

From these considerations we infer that the curve cuts the axis of y at two points equally distant from the origin; that it extends from these points both to the right and the left, gradually approaching the axis of x in each direction, and meeting that axis at points whose distances from the origin are a and $-a$; and, also, that there are no points of the curve farther from OY than the points in which it cuts OX.

Similarly, by examining equation (2), we see that the curve is limited in the direction of the axis of y by two points at distances b and $-b$ from the origin, lying wholly between these points.

Since the ellipse is symmetrical with respect to each axis, it must be symmetrical with respect to their intersection, the origin or centre of the curve; therefore every chord drawn through the centre is bisected at that point. It is easy to prove this analytically; for the equation of any line through the origin is, by § 25,

$$y = \lambda x, \quad \ldots\ldots\ldots\ldots (3)$$

and the equation of the ellipse is

$$\frac{x^2}{a^2} + \frac{y^2}{b^2} = 1. \quad \ldots\ldots\ldots\ldots (4)$$

Now if (x', y') is one point of intersection of the loci of (3) and (4), it is evident, from the form of these equations, that $(-x', -y')$ will also satisfy both and be the other point of intersection. Since the distance of each of these points from the origin is $\sqrt{x'^2 + y'^2}$, the origin or centre must be the middle point of each chord drawn through it.

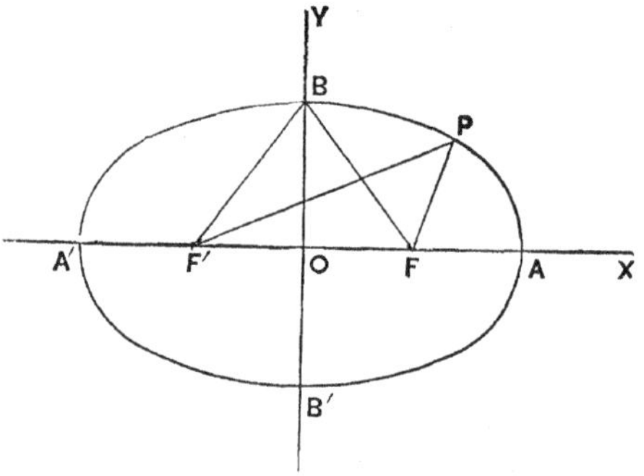

In the figure, A, A', B, B' are at distances a, $-a$, b, $-b$ from O respectively. The ellipse must cut the axes at these points, and will be inscribed in the rectangle whose sides are parallel to the axes and pass through the points A, A', B, B'. Its form, which is shown in the figure, may be determined with more precision by plotting a number of points, or by the following method based on the definition of the curve.

Let the extremities of a flexible string, whose length is $2a$, be fixed at the foci, F' and F; let P, the point of a pencil, move in the plane, keeping the string tight; its path must be the ellipse sought.

The *transverse*, or *major*, *axis* of an ellipse is that part of the straight line through the foci, contained by the curve; it is represented in the figure by $A' A$.

The *conjugate*, or *minor*, *axis* is that part of the perpendicular to the major axis, through the centre, which is included by the curve; it is shown in the figure as $B' B$.

The student has probably noticed that a, the half-sum of the focal radii, is equal to the semi-major axis of the ellipse, and that b is the semi-minor axis. In § 63 we introduced b as an abbreviation for $\sqrt{a^2 - c^2}$; hence $b < a$. Thus we see the reason for the terms major axis and minor axis.

These results may be obtained directly from the figure; for by the definition of the curve,

$$F' A + F A = 2a,$$

and, therefore, since $A' F'$ equals $F A$,

$$A' A = 2a \text{ and } O A = a;$$

also, since

$$F' B + F B = 2a,$$

and $F' B$ equals $F B$, we have, from the right triangle $O F B$,

$$O B = \sqrt{a^2 - c^2} = b.$$

The *eccentricity* of an ellipse is the ratio of c to a; this will always be less than unity [v. (1), § 63].

EXAMPLES.

(1.) The distance between the foci of an ellipse is 6, and the sum of the focal radii of any point is 10; find its equation.

Ans. $\frac{x^2}{25} + \frac{y^2}{16} = 1.$

(2.) Determine the semi-axes of the ellipse

$$x^2 + 2y^2 = 8.$$

This may be put in the form of equation (7), § 63, by dividing each term by 8, the constant term. This gives

$$\frac{x^2}{8} + \frac{y^2}{4} = 1,$$

which is the equation of an ellipse for which

$$a^2 = 8 \text{ and } b^2 = 4.$$

The required semi-axes are, therefore,

$$a = 2\sqrt{2} \text{ and } b = 2.$$

(3.) What are the semi-axes of the ellipse

$$16x^2 + 25y^2 = 1600?$$

Letting e represent the eccentricity of the ellipse, show that $e = \frac{3}{5}$.

(4.) Determine a, b, and e for the ellipse

$$\frac{x^2}{169} + \frac{y^2}{25} = 1.$$

(5.) The equation of a chord of the ellipse

$$x^2 + 7y^2 - 16 = 0,$$

is

$$x - 7y + 4 = 0;$$

find the length of the chord.

Ans. $\sqrt{50}$.

(6.) The *latus rectum*, or *parameter*, of a conic is the breadth of the curve through a focus; it is, therefore, equal to twice an ordinate erected at a focus.

Show that the latus rectum of an ellipse is equal to $\frac{2b^2}{a}$.

(7.) Prove that the minor axis of an ellipse is a mean proportional between the major axis and latus rectum.

(8.) Find the equations of the focal radii drawn to a point P_1 of the ellipse.

Ans. $\begin{cases} y_1 x - (x_1 + c)y + cy_1 = 0, \\ y_1 x - (x_1 - c)y - cy_1 = 0. \end{cases}$

(9.) Find the points of intersection of the circle and ellipse whose equations are

$$x^2 + y^2 = 10,\ x^2 + 7y^2 = 16.$$

Ans. (3, 1), (3, −1), (−3, 1), (−3, −1).

(10.) Find the points of intersection and the length of the common chord for the curves whose equations are

$$x^2 + y^2 + 20x + 50 = 0, \quad x^2 + 7y^2 - 16 = 0.$$

Ans. (−3, 1), (−3, −1), 2.

(11.) Prove that the semi-major axis of an ellipse is a mean proportional between the intercepts on that axis of the lines which join the extremities of the minor axis with any point of the curve.

68. It is worth while for us to notice that the circle is merely the special case of the ellipse in which the two foci coincide, for in this case $c = 0$, $b = a$, and the equation of the ellipse reduces to the familiar form of the equation of a circle.

It may be noticed that the eccentricity e of the circle is zero.

69. To investigate the form of the hyperbola, we first find its intercepts on the axes. Obtaining these from the equation (2) of § 64, we find the intercepts on the axis of x to be a and $-a$; those on the axis of y to be $b\sqrt{-1}$ and $-b\sqrt{-1}$, which are imaginary. The hyperbola, therefore, does not cut the axis of y, but since it has points on each side of that axis, it must comprise at least two parts or *branches*.

Solving equation (2), § 64, for y and x, we have

$$y = \pm \frac{b}{a}\sqrt{x^2 - a^2}, \qquad (1)$$

$$x = \pm \frac{a}{b}\sqrt{b^2 + y^2}, \qquad (2)$$

equations which show that the curve is symmetrical with respect to each axis.

From equation (1), we see that the values of y are imaginary when x is less than a, null when x equals a, and real when x is greater than a, in numerical value.

These facts show that there are no points of the curve, either above or below OX, between two points at distances a and $-a$ from the origin, while beyond these points the curve extends indefinitely in each direction of that axis.

We learn from (2), that all values of y give real values of x; consequently we infer that the curve runs without interruption in the direction of the axis of y, extending indefinitely in the positive and negative directions.

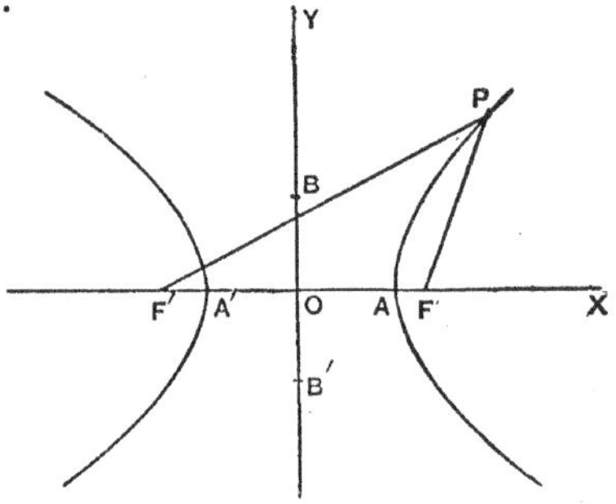

In either (1) or (2), we observe that x and y must increase together; therefore, the curve recedes from both axes at the same time.

We may show, as for the ellipse, that every chord drawn through the centre of an hyperbola is bisected by the centre.

The curve represented in the figure agrees with these results. It may be constructed by finding a number of its points, or, by a method* based on the definition of the curve, as follows:—

A pencil P is tied to a point near the middle of a piece of string. Two pegs are driven into the drawing-board at the points F and F', and the string is looped around these pegs in the manner indicated in the figure. The two parts of the string are grasped in one hand at H and drawn away from F'. The pencil P will then describe an arc of an hyperbola with foci at F and F', for the two distances FP and $F'P$ are evidently decreased by the same amounts, so that their differ-

* See papers by Ransom and Huntington in the Annals of Mathematics, Second Series, Vol. III, p. 164, and Vol. IV, p. 50.

ence remains constant. The other branch may be generated in the same manner by placing the pencil P at some

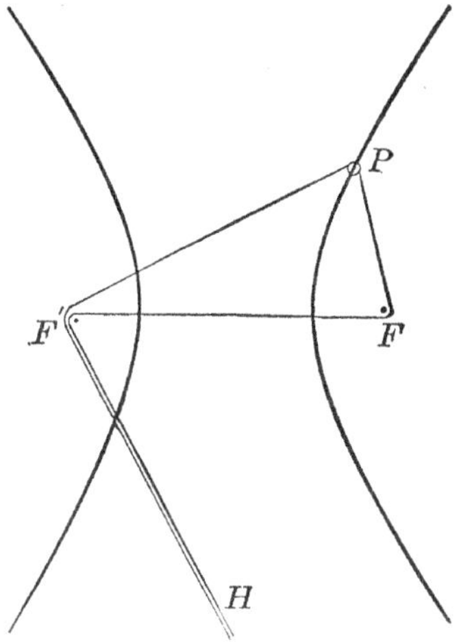

point of the other branch before drawing the string tight at H.

In the equation of the hyperbola, § 64, we used b to represent $\sqrt{c^2 - a^2}$; it will not, as for the ellipse, be the length of the intercepts on $O\,Y$, for the hyperbola does not cut $O\,Y$. We shall, however, call that portion of the axis of y contained between points at distances b and $-b$ from the origin, or centre, the *conjugate axis* of the hyperbola; its relation to the curve will be seen hereafter.

The definitions of the *transverse axis* and *eccentricity* are the same for the hyperbola as for the ellipse.

The transverse axis will lie between the two branches of the curve; the eccentricity must be greater than unity, because c is greater than a (v. § 64), and, for the same reason, the foci F' and F lie one within each branch of the hyperbola.

EXAMPLES.

(1.) The distance between the foci of an hyperbola is 10, and the difference between the focal radii of any point is 8; find its equation.

Ans. $\frac{x^2}{16} - \frac{y^2}{9} = 1.$

(2.) Determine the semi-axes and eccentricity of the hyperbola

$$9x^2 - 16y^2 = 576.$$

Using the same method as in Ex. 2, § **67**, we have

$$a = 8, \ b = 6, \ e = \frac{5}{4}.$$

(3.) The equation of a chord of the hyperbola

$$4x^2 - 9y^2 - 64 = 0$$

is

$$2x - 9y - 8 = 0;$$

find the length of the chord, and determine the semi-axes and eccentricity of the hyperbola.

Ans. Length of chord is $\sqrt{85}$; $e = \frac{\sqrt{13}}{3}$.

(4.) Show that the latus rectum of the hyperbola is $\frac{2b^2}{a}$; and state the theorem for the hyperbola analogous to that of § **67**, Ex. 7.

(5.) Show that the equations of the focal radii drawn to a point P_1 of the hyperbola are the same as the equations of like lines for the ellipse.

(6.) Find the points in which the following curves meet:—

$$\frac{x^2}{36} + \frac{5y^2}{81} = 1, \quad \frac{x^2}{9} - \frac{7y^2}{81} = 1.$$

Ans. $(4, \pm 3), (-4, \pm 3)$.

70. We will now consider the equation of the hyperbola in polar coördinates, as in this way a good deal of light will be thrown on the shape of the curve.

Using the formulæ of § 43, we have as the equation of the hyperbola

$$\frac{r^2 \cos^2 \phi}{a^2} - \frac{r^2 \sin^2 \phi}{b^2} = 1;$$

accordingly

$$r = \pm \frac{1}{\sqrt{\frac{\cos^2 \phi}{a^2} - \frac{\sin^2 \phi}{b^2}}}$$

$$= \pm \frac{b}{\cos \phi \sqrt{\frac{b^2}{a^2} - \tan^2 \phi}}.$$

Using the upper sign here and starting with the value $\phi = 0$, we have $r = a$. As ϕ increases, the denominator of r decreases, and therefore r itself increases. When ϕ has the value ϕ_0 for which

$$\tan \phi_0 = \frac{b}{a},$$

the denominator of r vanishes. Accordingly as ϕ increases and approaches the value ϕ_0 as a limit, r increases indefinitely. When ϕ is greater than ϕ_0 the expression under the radical sign is negative and r is therefore imaginary. Thus we see that in the first quadrant the hyperbola lies wholly below the line through the origin with slope $\frac{b}{a}$. In a similar manner we could discuss the shape of the curve in each of the other quadrants, or we can even more easily deduce the result from the symmetry of the curve. By either method we find that the hyperbola lies wholly between the two lines through the origin with slopes $\pm \frac{b}{a}$, while any line through the origin cuts the curve if it makes with the axis of X a smaller angle than these lines do. These two lines

$$y = \pm \frac{b}{a} x$$

are called the *asymptotes* of the hyperbola.

71. The most remarkable property of the asymptotes is that the hyperbola approaches them indefinitely as it is extended from the centre, but never reaches them. We may prove this property as follows: —

Draw corresponding ordinates for the hyperbola and the asymptote $O L$. These are represented in the figure by $M_1 P_1 = y_1$ and $M_1 P_1' = y_1'$. The equation of $O L$, the asymptote in the first and third quadrants, is

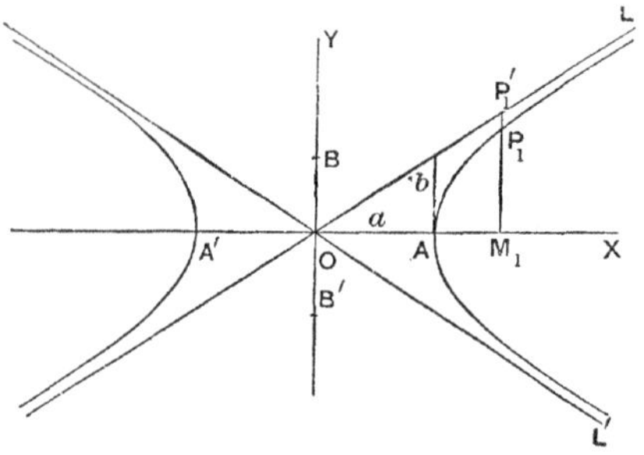

$$y = \frac{b}{a} x.$$

From this equation and that of the hyperbola, we obtain

$$y_1^2 = \frac{b^2}{a^2} (x_1^2 - a^2),$$

$$y_1'^2 = \frac{b^2}{a^2} x_1^2;$$

therefore

$$y_1'^2 - y_1^2 = \frac{b^2}{a^2} a^2 = b^2;$$

from which we may write

$$(y_1' - y_1)(y_1' + y_1) = b^2,$$

or

$$P_1 P_1' = y_1' - y_1 = \frac{b^2}{y_1' + y_1}.$$

Now as P_1 moves away from the centre, x_1 increases indefinitely, and the equations of the hyperbola and asymptote show that the accompanying values of y_1 and y_1' must also increase indefinitely. The value of $P_1 P_1'$ must therefore approach zero as its limit, for the numerator, b^2, is constant, while the denominator increases indefinitely. The distance of P_1 from the asymptote, measured on a perpendicular, is equal to $P_1 P_1'$ multiplied by a constant quantity, the sine of $P_1 P_1' O$; therefore this distance varies with $P_1 P_1'$, and approaches zero together with it. This proves the theorem.

It may be shown in like manner that a point below the axis of x, or one on the other branch of the hyperbola, approaches one of the asymptotes as it recedes from the centre, but can never really reach it.

EXAMPLES.

(1.) Find the equations of the asymptotes of the hyperbola

$$x^2 - 4y^2 = 36.$$

Ans. $x - 2y = 0$, $x + 2y = 0$.

(2.) Show that the distances of either focus from the asymptotes equal half the conjugate axis.

(3.) Discuss the shape of the ellipse by means of polar coördinates.

(4.) Prove that if from any point on an hyperbola a line is drawn parallel to either axis, the product of the parts cut off between the point and the asymptotes equals the square of half that axis.

72. Since in the hyperbola c is greater than a, and $b^2 = c^2 - a^2$, b may be less than a, equal to a, or greater than a, according as a is greater than, equal to, or less than $\frac{c}{\sqrt{2}}$.

When we make b equal to a, the equation of the hyperbola becomes

$$x^2 - y^2 = a^2,$$

and the curve is called an *equilateral hyperbola.*

The asymptotes of the equilateral hyperbola are

$$y = \pm x,$$

and since these lines are perpendicular to each other, this hyperbola is sometimes called the *rectangular* hyperbola.

73. To investigate the form of the parabola, we will discuss the equation found in § 66.

The intercepts of this curve are all 0. If we solve the equation for y and x, we have

$$y = \pm \sqrt{2\,m\,x}, \quad \ldots \ldots \ldots \ldots (1)$$

and

$$x = \frac{y^2}{2\,m}. \quad \ldots \ldots \ldots \ldots (2)$$

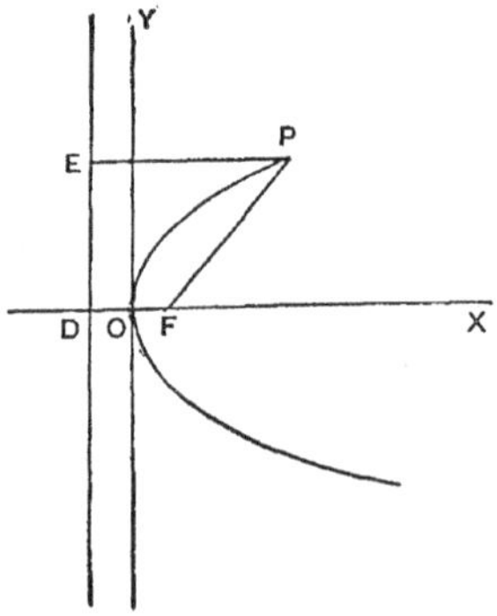

We learn from equation (1) that the curve is symmetrical with respect to OX. Also, if m is positive, all positive values of x give real values of y, but negative values of x make y imaginary; in this case the curve must lie wholly on the right of OY. In like manner we may show that when m is negative, the curve is wholly to the left of OY.

From (2) we see that all values of y

give real values of x; consequently the curve is continuous and indefinite in extent, in the direction of the axis of y.

Either (1) or (2) shows that x and y must increase together; therefore the curve recedes from both axes at the same time.

The parabola represented in the figure agrees with these results. It may be constructed by points, or by the following method based upon the definition of the curve: —

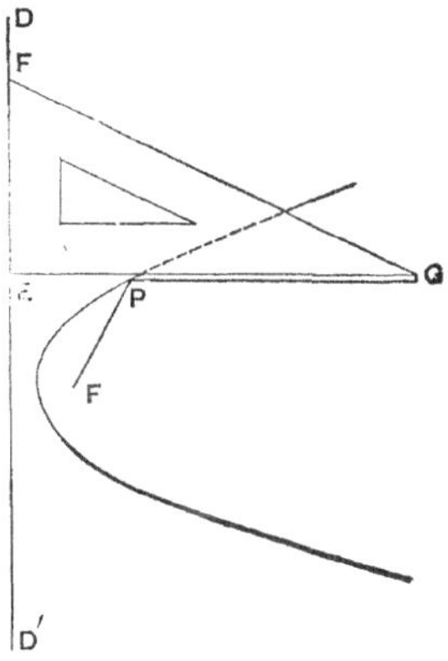

Let the leg $E F$, of the right-angled triangular ruler $E F G$ slide along the directrix $D' D$; fix one end of a string, equal to $E G$, at the focus F, and attach the other end to the ruler at G. A pencil point P, which presses the string tightly against $E G$, must move on a parabola.

The straight line through the focus and perpendicular to the directrix is called the *axis* of the parabola, and the point in which it cuts the curve is called the *vertex* of the parabola. The equation of the parabola found in § 66, is called the equation referred to the *vertex* and *axis*, or simply, to the *vertex*.

EXAMPLES.

(1.) What is the equation, referred to the vertex, of the parabola whose focus is at a distance 7 from its directrix? What is the equation of the same curve referred to its axis and directrix?

Ans. $y^2 = 14x$; $y^2 - 14x + 49 = 0$.

(2.) Show that the second locus constructed in § 17 is a parabola; how far is its focus from the directrix? how far from the vertex?

(3.) Find the latus rectum of the parabola.

Ans. $2m$.

(4.) Find the equations of the chords which join the vertex to the extremities of the latus rectum, the vertex being the origin.

Ans. $y = 2x, \quad y + 2x = 0.$

(5.) What is the equation of the circle which passes through the three points mentioned in the last example?

Ans. $2x^2 + 2y^2 = 5mx.$

(6.) The equations of two curves are

$$3x^2 + 4y^2 = 48 \text{ and } 9x - 2y^2 = 0.$$

What are the curves, and where do they intersect?

74. We will next find the equations of a tangent and a normal to an ellipse at a point P_1 of the curve.

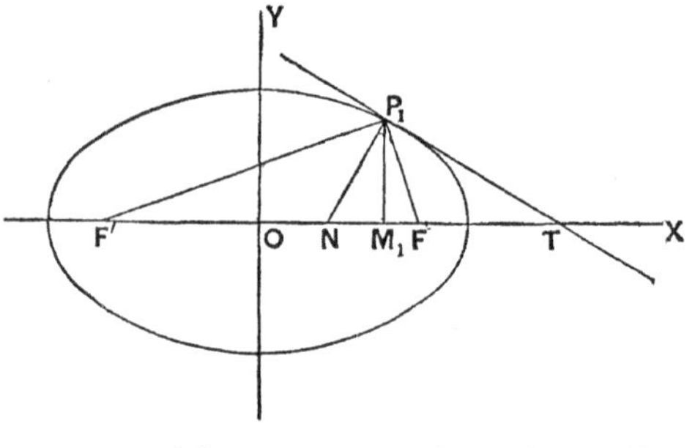

In the figure, $P_1 T$ is the tangent, and $P_1 N$ the normal. The coördinates of P_1 are $x_1 = O M_1$ and $y_1 = M_1 P_1$.

Using the same method as in § 49, we have, corresponding to equation (4) of that section, the following equations, obtained from equation (6), § 63:—

$$b^2 x_1^2 + a^2 y_1^2 = a^2 b^2, \quad \dots\dots\dots\dots \quad (1)$$

$$b^2(x_1 + h)^2 + a^2(y_1 + k)^2 = a^2 b^2. \quad \dots\dots\dots \quad (2)$$

Subtracting (1) from (2), we may write

$$2 b^2 h x_1 + b^2 h^2 + 2 a^2 k y_1 + k^2 a^2 = 0,$$

from which we obtain

$$\lambda_s = \frac{k}{h} = -\frac{b^2 (2x_1 + h)}{a^2 (2y_1 + k)} \quad \ldots\ldots\ldots \quad (3)$$

Taking the limit of this as k and h both approach zero, we have

$$\lambda_t = -\frac{b^2 x_1}{a^2 y_1}, \quad \ldots\ldots\ldots\ldots\ldots \quad (4)$$

and

$$\lambda_n = -\frac{1}{\lambda_t} = \frac{a^2 y_1}{b^2 x_1}. \quad \ldots\ldots\ldots\ldots \quad (5)$$

The equation of the tangent is, therefore,

$$y - y_1 = -\frac{b^2 x_1}{a^2 y_1}(x - x_1), \quad \ldots\ldots\ldots \quad (6)$$

and of the normal

$$y - y_1 = \frac{a^2 y_1}{b^2 x_1}(x - x_1) \quad \ldots\ldots\ldots\ldots \quad (7)$$

Clearing (6) of fractions, and transposing, we have

$$b^2 x_1 x + a^2 y_1 y = b^2 x_1^2 + a^2 y_1^2,$$

which, by equation (1), may be written

$$b^2 x_1 x + a^2 y_1 y = a^2 b^2, \quad \ldots\ldots\ldots\ldots \quad (8)$$

or

$$\frac{x_1 x}{a^2} + \frac{y_1 y}{b^2} = 1, \quad \ldots\ldots\ldots\ldots\ldots \quad (9)$$

the usual form of the equation of the tangent.

Clearing equation (7) of fractions, we may write it as follows: —

$$a^2 y_1 x \quad b^2 x_1 y - (a^2 - b^2) x_1 y_1 = 0, \quad \ldots\ldots \quad (10)$$

a form which may readily be obtained from (8) by the method of § 34.

EXAMPLES.

(1.) Find the equations of tangent and normal at (1, 2) to the ellipse

$$3x^2 + 4y^2 = 19.$$

Since the coördinates of the point satisfy the equation of the curve, the point lies on the ellipse, and, finding the semi-axes, we may write the equations of tangent and normal as follows:—

$$\frac{x}{\frac{19}{3}} + \frac{2y}{\frac{19}{4}} = 1,$$

or

$$3x + 8y = 19,$$

and

$$8x - 3y = 2.$$

(2.) Find the equations of tangents and normals to the ellipse

$$\frac{x^2}{12} + \frac{y^2}{6} = 1,$$

at all points of the curve whose abscissas are numerically equal to 2.

(3.) Calculate the intercepts on the axis of x of the tangent and normal drawn at a point P_1 of the ellipse.

$$Ans.\ \frac{a^2}{x_1},\ \frac{a^2 - b^2}{a^2}x_1 \text{ or } e^2 x_1.$$

(4.) Show that the lengths of the sub-tangent and sub-normal for a point P_1 of the ellipse are $\frac{a^2 - x_1^2}{x_1}$ and $\frac{b^2}{a^2}x_1$, respectively. (v. § 49, Ex. 12, and the figure of this section, where $M_1 T$ and $N M_1$ are the required lines).

(5.) By means of these results, prove that if a series of ellipses have the same major axis, tangents drawn to them at corresponding points (i. e., points which have the same abscissa) meet on the axis of x.

(6.) Find those points on an ellipse where the tangent is equally inclined to the axes.

$$Ans. \left(\pm\frac{a^2}{\sqrt{a^2+b^2}}, \pm\frac{b^2}{\sqrt{a^2+b^2}}\right)$$

) Find the equations of the tangents to the ellipse,

$$x^2 + 3y^2 = 3,$$

ı are inclined at an angle of 45° to the major axis.

Ans. $x - y + 2 = 0,\ x - y - 2 = 0.$

) Solve the last example when 45° is replaced by $\frac{3}{4}\pi$.

Ans. $x + y - 2 = 0,\ x + y + 2 = 0.$

) Find the equation of a tangent to the ellipse

$$\frac{x^2}{9a^2} + \frac{y^2}{4a^2} = 1.$$

e extremity of the latus rectum.

Ans. $\pm\sqrt{5}\,x \pm 3y = 9a$ represents the four tangents.

ɔ.) From the points where the circle on the major axis is sected by the minor axis produced, tangents are drawn e ellipse; find the points of contact.

Ans. The extremities of the latera recta.

.) Find the condition which must be fulfilled that the line

$$\frac{x}{m} + \frac{y}{n} = 1$$

be tangent to the ellipse

$$\frac{x^2}{a^2} + \frac{y^2}{b^2} = 1.$$

Ans. $\frac{a^2}{m^2} + \frac{b^2}{n^2} = 1.$

2.) Prove that if with the coördinates of any point of an se as semi-axes, a concentric ellipse be described with its in the same direction, the lines which join the extremities e axes of the first ellipse are tangent to the second.

. The method of finding the equations of a tangent and ıal to an hyperbola at a point of the curve is entirely analo-

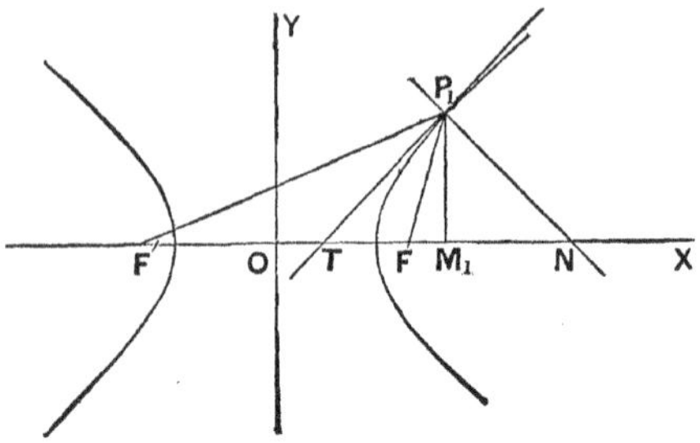

gous to that shown in the last section for the ellipse.

Show by this method that the equation of the tangent at P_1 is

$$\frac{x_1 x}{a^2} - \frac{y_1 y}{b^2} = 1 ; \quad . \ . \ . \ . \quad (1)$$

and of the normal.

$$(y - y_1) = -\frac{a^2 y_1}{b^2 x_1}(x - x_1), \quad . \ . \ . \ . \ . \ . \ . \ . \quad (2)$$

or

$$a^2 y_1 x + b^2 x_1 y - (a^2 + b^2) x_1 y_1 = 0. \quad . \ . \ . \ . \ . \ . \ . \quad (3)$$

NOTE. These equations evidently differ from the corresponding equations for the ellipse only in the sign of b^2, according to the principle stated in the note to § 64.

EXAMPLES.

(1.) Find the equations of tangent and normal to the hyperbola

$$2x^2 - 3y^2 = 12,$$

at the point $(3, \sqrt{2})$.

Ans. $\begin{cases} 2x - \sqrt{2}\, y = 4, \\ \sqrt{2}\, x + 2y = 5\sqrt{2}. \end{cases}$

(2.) Find the equations of tangents and normals to the hyperbola

$$\frac{x^2}{9} - \frac{y^2}{3} = 1,$$

at all points of the curve whose ordinates are numerically equal to 3.

(3.) Show that the expressions for the intercepts on OX of a tangent and normal at a point P_1 of the hyperbola are the same as for the analogous lines in the ellipse.

(4.) What is the property of hyperbolas analogous to that stated in § 74, Ex. 5?

(5.) Show that the lengths of the sub-tangent and sub-normal of the hyperbola have the same expressions as for the ellipse, because the change in the sign of b^2 does not affect the numerical value of the sub-normal.

(6.) Find those points on the hyperbola at which a tangent has an inclination of 45°; of 135°.

$$Ans. \begin{cases} \left(\frac{a^2}{\sqrt{a^2-b^2}}, \frac{b^2}{\sqrt{a^2-b^2}}\right), \left(-\frac{a^2}{\sqrt{a^2-b^2}}, -\frac{b^2}{\sqrt{a^2-b^2}}\right); \\ \left(\frac{a^2}{\sqrt{a^2-b^2}}, -\frac{b^2}{\sqrt{a^2-b^2}}\right), \left(-\frac{a^2}{\sqrt{a^2-b^2}}, \frac{b^2}{\sqrt{a^2-b^2}}\right). \end{cases}$$

(7.) Is the problem of the last example possible for every hyperbola? If not, what are the exceptions?

(8.) Find the equations of normals to the hyperbola

$$9x^2 - 25y^2 = 225,$$

which have an inclination of 45°.

(9.) Find the equations of tangents and normals to an hyperbola at the extremities of the parameters.

Ans. For the upper extremity of the right-hand latus rectum,

$$ex - y = a, \quad ax + aey = e(a^2 + b^2).$$

(10.) Find the condition which must be fulfilled that the line

$$\frac{x}{m} + \frac{y}{n} = 1$$

may touch the hyperbola

$$\frac{x^2}{a^2} - \frac{y^2}{b^2} = 1.$$

76. Let us find the equations of a tangent and a normal to a parabola at a point P_1 of the curve. Using the simplest form of the equation of a parabola

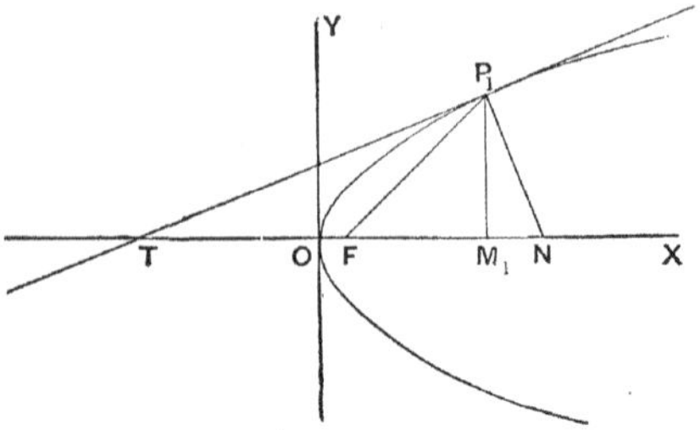

$$y^2 = 2\,m\,x \;.\;.\;.\; (1)$$

we have for the equations from which to find the value of λ_s,

$$y_1^2 = 2\,m\,x_1 \;.\;.\;.\; (2)$$

and

$$(y_1 + k)^2 = 2\,m\,(x_1 + h). \; (3)$$

From these we easily fird

$$\lambda_s = \frac{k}{h} = \frac{2\,m}{2\,y_1 + k}, \quad \ldots\ldots\ldots\ldots (4)$$

which gives

$$\lambda_t = \frac{m}{y_1}, \quad \ldots\ldots\ldots\ldots (5)$$

and

$$\lambda_n = -\frac{y_1}{m} \quad \ldots\ldots\ldots\ldots (6)$$

The required equations are, therefore,

$$y - y_1 = \frac{m}{y_1}(x - x_1), \ldots\ldots\ldots\ldots (7)$$

or

$$y_1\,y = m\,(x + x_1), \ldots\ldots\ldots\ldots (8)$$

and

$$y - y_1 = -\frac{y_1}{m}(x - x_1), \ldots\ldots\ldots\ldots (9)$$

or

$$y_1\,x + m\,y = (m + x_1)\,y_1, \ldots\ldots\ldots\ldots (10)$$

EXAMPLES.

(1.) Find the equations of tangent and normal to the parabola

$$y^2 = 4\,x,$$

at the point (1, 2).

Ans. $\begin{cases} x - y + 1 = 0, \\ x + y - 3 = 0. \end{cases}$

(2.) Obtain the equations of tangents and normals to the parabola

$$y^2 = 8x$$

at all points of the curve whose abscissas are equal to 8.

(3.) Show that the intercepts on OX of the tangent and normal at a point P_1 of the parabola

$$y^2 = 2mx$$

are $-x_1$ and $m + x_1$, respectively.

(4.) Find the lengths of the sub-tangent and sub-normal for the parabola; show that the first is bisected at the vertex, and that the second is constant.

(5.) Prove that the tangents at the extremities of the latus rectum are equally inclined to the axes of x and y, and meet the directrix on the axis of x (v. § 67, Ex. 6).

(6.) Form the equations of normals to the parabola at the extremities of the latus rectum; show that they are mutually perpendicular and intersect on the axis of the curve. What kind of quarilateral is bounded by these normals and the tangents mentioned in the last example? What is its area?

$$Ans. \begin{cases} 2(x + y) = 3m, \\ 2(x - y) = 3m, \\ \text{Area} = 2m^2. \end{cases}$$

(7.) Show that if the focus of a parabola is the origin and the axis of the curve, the axis of x, the equation of the tangent at P_1, a point of the curve, is

$$y_1 y = m(x + x_1 + m).$$

(8.) With the same axes as in the last example, find the tangents and normals at the extremities of the latus rectum.

$$Ans. \begin{cases} x \mp y + m = 0, \\ x \pm y - m = 0. \end{cases}$$

(9.) Write the equations of the directrix and latus rectum, referred to the vertex.

(10.) Show that the tangent at any point of the parabola meets the directrix and latus rectum produced at points equally distant from the focus, the common distance being $\frac{m^2 + y_1^2}{2 y_1}$.

(11.) Find the condition that the line

$$\frac{x}{a} + \frac{y}{b} = 1$$

may touch the parabola

$$y^2 = 2 m x.$$

Ans. $2 b^2 + a m = 0$.

77. Instead of finding the tangent to the ellipse

$$\frac{x^2}{a^2} + \frac{y^2}{b^2} = 1 \quad \ldots \ldots \ldots \ldots \ldots (1)$$

at the given point (x_1, y_1), it is sometimes necessary to find the tangent to this ellipse *which has a given slope* λ. It is obvious, when we consider the shape of the ellipse, that there are two such tangents. Their equations can be found as follows.

Denoting the coördinates of the point of contact of one of these tangents by (x_1, y_1), its equation may be written

$$\frac{x_1 x}{a^2} + \frac{y_1 y}{b^2} = 1, \quad \ldots \ldots \ldots \ldots (2)$$

or

$$y = -\frac{x_1}{y_1}\frac{b^2}{a^2} x + \frac{b^2}{y_1} \quad \ldots \ldots \ldots \ldots (3)$$

From (3) we have

$$\lambda = -\frac{x_1}{y_1}\frac{b^2}{a^2}, \quad \ldots \ldots \ldots \ldots \ldots (4)$$

and the equation of the tangent can be written

$$y = \lambda x + \frac{b^2}{y_1} \quad \ldots \ldots \ldots \ldots (5)$$

It remains merely to replace y_1 in this equation by its value in terms of λ. This value we find as follows. Squaring (4) and clearing of fractions, we have

$$b^4 x_1^2 - a^4 \lambda^2 y_1^2 = 0. \quad \ldots \ldots \ldots \quad (6)$$

On the other hand, since (x_1, y_1) is a point of the ellipse (1), it satisfies this equation, and we have

$$b^2 x_1^2 + a^2 y_1^2 = a^2 b^2. \quad \ldots \ldots \ldots \quad (7)$$

By multiplying (7) by b^2 and subtracting (6) from it we eliminate x_1 and get

$$a^2 (a^2 \lambda^2 + b^2) y_1^2 = a^2 b^4$$

or

$$y_1 = \frac{\pm b^2}{\sqrt{a^2 \lambda^2 + b^2}}.$$

Substituting this value of y_1 in (5), we get as the desired equation of the tangent

$$y = \lambda x \pm \sqrt{a^2 \lambda^2 + b^2}. \quad \ldots \ldots \ldots \quad (8)$$

The two signs here give us the two tangents with slope λ.

In the same way we find that the equations of the tangents to the hyperbola which have the slope λ are

$$y = \lambda x \pm \sqrt{a^2 \lambda^2 - b^2}. \quad \ldots \ldots \ldots \quad (9)$$

It should be noticed that this equation becomes imaginary when λ is numerically less that $\frac{b}{a}$; so that the hyperbola has no tangents which make a smaller angle with the transverse axis than that made by the asymptotes.

In the case of the parabola it is easily seen that there is only one tangent with a given slope λ. By using the method explained above, its equation is found to be

$$y = \lambda x + \frac{m}{2\lambda}. \quad \ldots \ldots \ldots \quad (10)$$

EXAMPLES.

(1.) Deduce the results of this section by starting from the equation $y = \lambda x + \beta$ and imposing the condition that the points in which this line cuts the conic should coincide.

(2.) A right angle moves so that both of its sides touch an ellipse. Find the locus of its vertex.

(3.) Solve the last problem if instead of an ellipse we have a parabola.

(4.) Find the equations of the tangents from the point (14, 1) to the ellipse $x^2 + 4y^2 = 100$.

Suggestion: Call the coördinates of the point of contact of one of these tangents (x_1, y_1), and then determine the values of these coördinates.

78. Let us find the lengths of the focal radii (r and r' in the figure of § 63) drawn to a point P_1 of the ellipse.

By § 10, we have

$$r^2 = (x_1 - c)^2 + y_1^2, \quad \dots \dots \dots \dots \quad (1)$$
$$r'^2 = (x_1 + c)^2 + y_1^2. \quad \dots \dots \dots \dots \quad (2)$$

Also we know that the coördinates of P_1 must satisfy the equation of the ellipse, and therefore

$$y_1^2 = \frac{b^2}{a^2}(a^2 - x_1^2), \quad \dots \dots \dots \dots \quad (3)$$

Substituting this value of y_1^2 in equation (1), we have after expanding and collecting,

$$r^2 = \frac{a^2 - b^2}{a^2} x_1^2 - 2\,c\,x_1 + c^2 + b^2 \quad \dots \dots \dots \quad (4)$$

Now $a^2 - b^2 = c^2$ and therefore $c^2 + b^2 = a^2$; also $e = \frac{c}{a}$ and consequently $c = a\,e$. We may therefore write equation (4) as follows:—

$$r^2 = e^2 x_1^2 - 2\,a\,e\,x_1 + a^2;$$

and

$$r = e\,x_1 - a, \text{ or } a - e\,x_1 \quad \dots \dots \dots \quad (5)$$

As we only wish the length of r, we shall use the second of these values, because the first is necessarily negative, e being less than unity and x_1 less than a.

In like manner show that

$$r' = a + e\,x_1 \quad \dots\dots\dots\dots \quad (6)$$

This may also be proved from (5) and the definition of the curve.

EXAMPLES.

(1.) Prove that if on any ellipse three points be taken whose abscissas are in arithmetical progression, the corresponding focal radii of these points are also in arithmetical progression.

(2.) Prove that if the ordinate of any point of an ellipse be extended to meet the tangent drawn at the extremity of the latus rectum, the length of the extended ordinate equals that of the corresponding focal radius of the point.

(3.) The ordinate of any point of an ellipse is extended to meet the circumscribed circle, and a tangent is drawn to the circle at its extremity; prove that either focus is equally distant from the point and the tangent.

79. We may easily find the segments, $F'N$ and NF, cut by the normal from the base of the triangle $F'P_1F$, in the figure of § 74, by adding and subtracting ON and c. They are

$$F'N = a\,e + e^2\,x_1,\ NF = a\,e - e^2\,x_1.$$

Comparing these values, we have

$$\frac{F'N}{NF} = \frac{a\,e + e^2\,x_1}{a\,e - e^2\,x_1} = \frac{a + e\,x_1}{a - e\,x_1} = \frac{r'}{r}.$$

Since the base is divided into segments proportional to the adjacent sides, we know, from Geometry, that *the angle between the focal radii drawn to any point is bisected by the normal at that point.*

The normal is the internal bisector of the angle between the focal radii, consequently the tangent perpendicular to the normal must bisect the supplementary external angle.

80. For the hyperbola we may show, as in § 78, that for the right-hand branch (when x_1 is positive) the focal radii are

$$r' = e\,x_1 + a, \text{ and } r = e\,x_1 - a\,; \quad \ldots\ldots\ldots \quad (1)$$

for the left-hand branch (where x_1 is negative)

$$r' = -\,e\,x_1 - a, \text{ and } r = a - e\,x_1 \quad \ldots\ldots\ldots \quad (2)$$

EXAMPLE.

Show that the theorems stated in the examples 1 and 2 of § 78 are true for the hyperbola.

81. The tangent to an hyperbola at a point P_1 will be found to cut the base of the triangle $F'\,P_1\,F$ (§ 75, Figure) into segments

$$F'\,T = a\,e + \frac{a^2}{x_1}, \quad T F = a\,e - \frac{a^2}{x_1}.$$

Comparing these segments, we have

$$\frac{F'\,T}{T F} = \frac{a\,e\,x_1 + a^2}{a\,e\,x_1 - a^2} = \frac{e\,x_1 + a}{e\,x_1 - a} = \frac{r'}{r}.$$

The tangent therefore bisects internally the angle between the focal radii of the hyperbola drawn to the point of tangency, and consequently the normal at that point bisects the supplementary external angle.

82. The results of §§ 79 and 81 show us that when an ellipse and hyperbola have the same foci, the curves cut each other at right angles. For, at the points of intersection, the curves have the same focal radii, the tangent to the hyperbola bisects the internal angle between these radii, and the tangent to the ellipse bisects the external angle. These tangents are therefore mutually perpendicular; and, since a tangent to a curve at any point has the direction of the curve at that point, we learn that *confocal conics intersect at right angles.*

83. It follows from the definition of the parabola, that the focal radius FP_1 (v. figure of § 76) drawn to a point P_1 of the curve is equal to $\frac{m}{2} + x_1$. Now we may easily show, by the results of Ex. 3, § 76, that TF is equal to $x_1 + \frac{m}{2}$; therefore, the triangle TFP_1 is isosceles, and *the tangent to a parabola at any point must make equal angles with the axis and the focal radius drawn to that point.*

From this it follows that if at any point of the curve the focal radius is drawn and a line parallel to the axis, the tangent and normal at that point bisect the external and internal angles between these lines.

84. If we describe a circle on the major axis of an ellipse as a diameter, corresponding ordinates of the ellipse and circle will have a constant ratio, $b : a$.

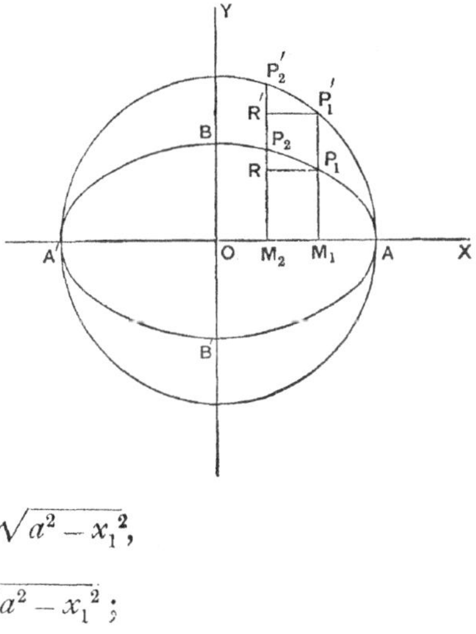

In the figure, y_1 $(= M_1 P_1)$ and y_1' $(= M_1 P_1')$ are corresponding ordinates, because they are drawn with the same abscissa x_1 $(= OM_1)$. From the equations of the ellipse and circle, we have for the values of these ordinates

$$y_1 = \frac{b}{a} \sqrt{a^2 - x_1^2},$$

$$y_1' = \sqrt{a^2 - x_1^2};$$

therefore, remembering that a and x_1 are by hypothesis the same in these equations,

$$\frac{y_1}{y_1'} = \frac{b}{a},$$

which was to be proved.

It may be shown in like manner that when a circle is inscribed in an ellipse, with the minor axis as a diameter, corresponding abscissas of ellipse and circle have the constant ratio, $a : b$.

By means of these properties we can construct an ellipse by shortening each ordinate of the circumscribed circle in the ratio $b : a$, or by lengthening each abscissa of the inscribed circle in the ratio $a : b$.

We can also draw a tangent at any point of an ellipse by drawing the tangent to the circumscribed circle at the corresponding point, and joining the point in which this tangent meets the major axis produced, with the point of the ellipse at which the tangent is to be drawn; for the circumscribed circle may be regarded as one of the series of ellipses mentioned in § 74, Ex. 5.

EXAMPLE.

Two tangents to the circle whose equation is

$$x^2 + y^2 = 100$$

meet in the point (14, 2); find the equations of tangents to the corresponding points of the ellipse

$$x^2 + 4y^2 = 100.$$

Ans. $3x + 8y = 50$, $2x - 3y = 25$.

85. By means of the theorem of § 84, we may calculate the area of an ellipse.

Let $A'A$, in the figure of § 84, be divided into any number of equal parts, and suppose M_1 and M_2 two of the dividing points. Erect corresponding ordinates at these points, and at their extremities draw parallels to $A' A$, forming pairs of *corresponding rectangles*, of which $M_1 R$ and $M_1 R'$ are one pair.

Now from the figure we see that

$$\frac{M_1 R}{M_1 R'} = \frac{\overline{M_1 M_2} \times y_1}{M_1 M_2 \times y_1'} = \frac{y_1}{y_1'} = \frac{b}{a}.$$

Similarly we may show that the area of any corresponding rectangles of ellipse and circle are in the ratio $b : a$; therefore the sum of the rectangles of the ellipse is to the corresponding sum for the circle as $b : a$. Now if we increase indefinitely the number of equal divisions of $A' A$, the sum of the rectangles of the ellipse approaches the half-area of the ellipse as a limit, and the rectangles of the circle approach the half-area of the circle as a limit; therefore, by the Theory of Limits, the area of the ellipse is to the area of the circle as $b : a$. Since the area of the circle is πa^2, the area of the ellipse must be $\pi a^2 \frac{b}{a}$, or $\pi a b$.

EXAMPLES.

(1.) Find the area of the ellipse whose equation is

$$x^2 + 4 y^2 = 100.$$

(2.) Find the ratio of the areas of the ellipses

$$9 x^2 + 16 y^2 = 144,$$
$$x^2 + 4 y^2 = 36,$$
$$4 x^2 + 9 y^2 = 144.$$

Ans. $2 : 3 : 4$.

86. Let us find the equation of an hyperbola whose foci lie on the axis of y, whose transverse axis is $2 b$, and whose conjugate axis is $2 a$.

We will first find the equation of an hyperbola, with its foci on $O Y$, with $2 a$ for its transverse, and $2 b$ for its conjugate, axis. This may be obtained from the equation of § 64, by turning the axes through an angle of 90°. Using the formulas for transformation given in § 42, Cor., we have for the equation of this hyperbola

$$\frac{y^2}{a^2} - \frac{x^2}{b^2} = 1. \quad . \quad . \quad . \quad . \quad . \quad . \quad . \quad . \quad . \quad . \quad . \quad . \quad (1)$$

The transverse axis of the required hyperbola has the same length as the conjugate axis of this one, and *vice versa;* therefore we shall obtain the required equation by interchanging a and b in (1). This gives

$$\frac{y^2}{b^2} - \frac{x^2}{a^2} = 1, \text{ or } \frac{x^2}{a^2} - \frac{y^2}{b^2} = -1, \quad . \quad . \quad . \quad . \quad . \quad . \quad (2)$$

which must be the equation sought.

This hyperbola is said to be conjugate to the one whose equation was found, and the two are closely connected. We may therefore define *conjugate hyperbolas* as those which have the same axes, the transverse axis of one being the conjugate axis of the other, and the conjugate axis of the one the transverse axis of the other.

EXAMPLES.

(1.) Find the equation of the hyperbola conjugate to

$$x^2 - 4y^2 = 36,$$

and determine its foci, axes, and eccentricity.

Ans. $4y^2 - x^2 = 36$, $(0, \pm 3\sqrt{5})$, 3, 6, $\sqrt{5}$,

are the equation, foci, semi-transverse axis, semi-conjugate axis, and eccentricity, respectively.

(2.) Find the equations of tangent and normal to the hyperbola conjugate to

$$\frac{x^2}{a^2} - \frac{y^2}{b^2} = 1,$$

at a point P_1 of the curve

Ans. $$\begin{cases} \dfrac{y_1 y}{b^2} - \dfrac{x_1 x}{a^2} = 1 \text{ and} \\ b^2 x_1 y + a^2 y_1 x = (a^2 + b^2) x_1 y_1. \end{cases}$$

87. *Conjugate hyperbolas have the same asymptotes.*

For, turning the axes through 90°, and interchanging a and b in the equations of the asymptotes to the hyperbola

$$\frac{x^2}{a^2} - \frac{y^2}{b^2} = 1,$$

found in § 70, we have

$$x = \pm \frac{a}{b} y, \text{ or } y = \pm \frac{b}{a} x, \qquad (1)$$

for the equations of the asymptotes to the hyperbola

$$\frac{y^2}{b^2} - \frac{x^2}{a^2} = 1.$$

Comparing (1) with the equations of the asymptotes in § 70, we see that the equations are alike, which proves our proposition.

88. In the accompanying figure, we show a pair of conjugate hyperbolas with their common asymptotes OL and OL'.

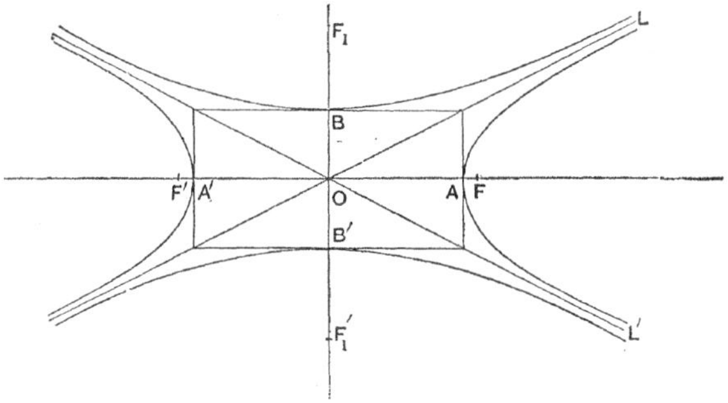

$A' A$ and $B' B$ are axes of both curves; the one with $A' A$ for its transverse axis has foci F and F'; the one which has $B B'$ for its transverse axis has foci F_1 and F_1'.

The reader may easily deduce the following properties from those already proved and the definitions which have been given:

The four foci lie on the circumference of a circle whose radius is c.

The rectangle which has its sides parallel and equal to the axes, and the common centre of the hyperbolas for its centre, lies wholly between the curves, having each side tangent to one of the branches of these curves, and the asymptotes of the hyperbolas for its diagonals.

An ellipse with the same axes as the hyperbolas will be inscribed in this rectangle and will touch the hyperbolas at A, A', B, B'.

89. The definitions of the conic sections which we gave at the beginning of this chapter fail to bring out the fact, which appears on closer study, that these three curves are merely different species of a single genus. It will therefore be of interest to give a new definition which covers equally the three cases of the ellipse, the hyperbola, and the parabola. We shall in this way gain a more comprehensive view of these curves and learn some new facts about them.

BOSCOVICH'S DEFINITION.—A Conic Section is the locus of a point which moves in a plane so that its distance from a fixed point of the plane bears a constant ratio to its distance from a fixed line in the plane.

The fixed point is called the *focus*, the fixed line the *directrix*, and the value of the constant ratio the *eccentricity* of the conic section.

We will now examine this definition and show that it is not in contradiction with the definitions previously laid down.

Let us take the directrix in the above definition as axis of Y and the perpendicular dropped upon it from the focus F as axis of X. We will call the eccentricity e, and the distance from the directrix to the focus m, so that the coördinates of F are $(m, 0)$. Denoting by (x, y) the coördinates of the moving point P which traces out the locus, we obtain at once as the equation which expresses the definition of the curve

$$\frac{\sqrt{(x-m)^2+y^2}}{x} = e \quad \ldots\ldots\ldots\ldots (1)$$

This equation, when simplified, takes the form

$$(1 - e^2)x^2 + y^2 - 2mx + m^2 = 0 \ldots\ldots (2)$$

This is the equation of the conic section.

If $e = 1$ this equation reduces to equation (3), § 65, and in fact it will be seen that in this case our present definition of a conic section is the same as our old definition of a parabola, and our use of the terms *focus* and *directrix* coincide in these cases. We have merely introduced the new term *eccentricity*, which up to this time we had not used in the case of the parabola, in such a way that the eccentricity of every parabola is 1.

Considering now the cases where $e \neq 1$, let us find the intercepts of our curve on the axis of X. This is done by letting $y = 0$ in (2):

$$(1 - e^2)x^2 - 2mx + m^2 = 0.$$

The roots of this equation are $\frac{m}{1 \pm e}$. Accordingly the curve (2) cuts the axis of X in the two points A, B:

$$\left(\frac{m}{1+e}, 0\right), \quad \left(\frac{m}{1-e}, 0\right). \ldots\ldots (3)$$

Denoting the distance between these two points by $2a$ we have

$$a = \frac{em}{1-e^2}. \ldots\ldots (4)$$

In order to simplify equation (2) let us now transform it to a new system of rectangular coördinates parallel to the old, with origin half way between the points (3), *i.e.* at the point C:

$$\left(\frac{m}{1-e^2}, 0\right).$$

The formulas for this transformation are (v. (3), § 41):

$$x = x' + \frac{m}{1 - e^2}, \quad y = y'.$$

Performing this transformation and dropping accents, equation (2) becomes

$$(1 - e^2)\,x^2 + y^2 - \frac{e^2 m^2}{1 - e^2} = 0.$$

If we introduce the quantity a by means of (4) this equation takes the form

$$\frac{x^2}{a^2} + \frac{y^2}{a^2(1 - e^2)} = 1. \quad \ldots\ldots\ldots \quad (5)$$

Let us examine first the case $e < 1$. In this case (5) represents an ellipse whose semi-axes are a and

$$b = a\sqrt{1 - e^2} = \frac{e\,m}{\sqrt{1 - e^2}}, \quad \ldots\ldots\ldots \quad (6)$$

this latter being obviously the semi-minor axis.

Moreover, we have

$$FC = OC - OF = \frac{m}{1 - e^2} - m = \frac{e^2 m}{1 - e^2} = a\,e = \sqrt{a^2 - b^2}. \quad (7)$$

Accordingly F, which we have called the focus of the conic, is really one of the foci of the ellipse according to our earlier definition.

Finally the eccentricity of the ellipse was defined on p. 122 as

$$\frac{c}{a} = \frac{FC}{a} = \frac{a\,e}{a} = e,$$

so that our new definition of the eccentricity coincides with the old.

The conception of a directrix for the ellipse is something we did not have before. Since the ellipse is symmetrical with regard to the minor axis, it is clear that, besides the focus and directrix with which we started, there is a second focus and a second directrix situated the same distance to the right of the minor axis as the first were to the left. The distance from the directrix to the minor axis is

$$OC = \frac{m}{1 - e^2} = \frac{a}{e}.$$

Thus, when the centre of the ellipse is taken as origin, our two directrices are

$$x = \pm \frac{a}{e} \quad \ldots \ldots \ldots \ldots \ldots \quad (8)$$

Turning now to the case $e > 1$ we see that (5) represents a hyperbola whose semi-transverse axis is a, while its semi-conjugate axis is

$$b = a\sqrt{e^2 - 1} \quad \ldots \ldots \ldots \ldots \quad (9)$$

Moreover, we have

$$FC = OC - OF = \frac{e^2 m}{1 - e^2} = ae = \sqrt{a^2 + b^2}, \quad (10)$$

so that F is really a focus of the hyperbola according to our earlier definition.

As in the case of the ellipse we see that our new definition of the eccentricity coincides with the old.

We also see that the hyperbola has two directrices whose equations are (8).

The ellipse and hyperbola are sometimes called the *central conics* in distinction to the parabola, which has no centre.

90. From the point of view of the last section the three kinds of conic sections are distinguished from one another by the value of the eccentricity, and we have

An Ellipse when $e < 1$,
A Parabola when $e = 1$,
An Hyperbola when $e > 1$.

Thus we see that the parabola may be regarded as a form of curve intermediate between the ellipse and the hyperbola. If in equation (2) we keep m, the distance from the directrix to the focus, unchanged and allow e, starting with a value less than 1, to increase, we have an ellipse whose major axis (v. (4)) and whose minor axis (v. (6)) are both increasing. If we allow e to increase towards the value 1 as its limit, these axes increase indefinitely, and the ellipse approaches a parabola as its limiting form. On the other hand, if e, starting with a value greater than 1, decreases towards 1 as its limit, the axes of the hyperbola represented by (2) increase indefinitely. One branch of the hyperbola, therefore—the left-hand one if m is positive—moves farther and farther away, while the other branch approaches a parabola as its limiting form.

91. It can easily be seen that *every* conic except a circle can be obtained by the method of § 89. This is clearly so for parabolas, since in this case the method of generating the conic given in § 89 is practically the same as that given in § 62. In the case of an ellipse or hyperbola the conic is completely determined by its semi-axes a and b. These in turn determine the eccentricity e, and e and a determine m by (4) except in the case of a circle, for which $e = 0$, while a is in general not zero, so that (4) involves a contradiction. The fact that the circle is not included under Boscovich's definition of a conic is easily understood when we notice that

the circle is the limiting form of an ellipse as the eccentricity approaches zero. Accordingly by (8), § 89, the directrices of the ellipse move off indefinitely, so that the circle, having no directrices, cannot come under a definition which involves the use of a directrix.

EXAMPLES.

(1.) Find the equation of the tangent to the conic (2) at the point (x_1, y_1).

(2.) Prove that tangents drawn to any conic at the extremities of a focal chord intersect on the directrix.

(3.) A point moves so that its distance from a fixed point is always equal to its distance from a fixed circle. Prove that its locus is a conic having the fixed point as focus.

REMARK.—The fixed circle of this example is sometimes called the *director circle* of the conic.

(4.) Show that every central conic has two director circles (v. Ex. 3), and find their equations.

CHAPTER VIII.

DIAMETERS. POLES AND POLARS.

92. According to the definition of p. 89, the diameter of a conic section is the locus of the middle points of a set of parallel chords.

Beginning with the ellipse, let us find the locus of the middle points of a set of parallel chords whose equations are

$$y = \lambda_1 x + b_1 \quad \ldots\ldots\ldots\ldots \quad (1)$$
$$y = \lambda_1 x + b_2 \quad \ldots\ldots\ldots\ldots \quad (2)$$

etc.

The extremities of the first chord may be found by combining its equation with that of the ellipse.

Using the form of the equation given in § 63, equation (6), and substituting for y^2 its value from equation (1), we have

$$(b^2 + a^2 \lambda_1^2) x^2 + 2 a^2 b_1 \lambda_1 x + a^2 (b_1^2 - b^2) = 0.$$

The roots of this equation are the abscissas of the extremities of the chord, and therefore, by § 11, and the foot-note on p. 90, the abscissa x' of the middle point of the chord is

$$x' = -\frac{a^2 b_1 \lambda_1}{b^2 + a^2 \lambda_1^2} \quad \ldots\ldots\ldots\ldots \quad (3)$$

The coördinates of P', the middle point of chord (1), must of course satisfy equation (1), and therefore

$$y' = -\frac{a^2 b_1 \lambda_1^2}{b^2 + a^2 \lambda_1^2} + b_1,$$

which becomes, after reducing,

$$y' = \frac{b^2 b_1}{b^2 + a^2 \lambda_1^2} \quad \ldots\ldots\ldots\ldots \quad (4)$$

Dividing y' by x', we obtain from (4) and (3) a necessary relation between them which is independent of b_1, the intercept of the first chord on the axis of y. It is

$$\frac{y'}{x'} = -\frac{b^2}{a^2 \lambda_1} \quad \ldots \ldots \ldots \ldots \quad (5)$$

This relation must exist between the coördinates of the middle points of each of the set of parallel chords, and is therefore the equation of a diameter. Dropping the accents we may write it

$$y = -\frac{b^2}{a^2 \lambda_1} x, \quad \ldots \ldots \ldots \ldots \quad (6)$$

which is in the form

$$y = \lambda x,$$

where

$$\lambda = -\frac{b^2}{a^2 \lambda_1} \quad \ldots \ldots \ldots \ldots \quad (7)$$

From the form of the equation, we know that the diameter passes through the origin or centre, but it is not, as for the circle, perpendicular to the chords which it bisects, for the slopes of chord and diameter do not satisfy the test of perpendicularity.

It may be shown in the same manner as for the circle (v. § 52) that every chord through the centre is a diameter.

93. Using λ_2 to represent the slope of that diameter which bisects chords with a slope λ_1, let us find the slope of the diameter which bisects all chords whose slopes are λ_2.

Calling the required slope λ, we have by equation (7) of the last section

$$\lambda = -\frac{b^2}{a^2 \lambda_2}, \quad \ldots \ldots \ldots \ldots \quad (1)$$

where

$$\lambda_2 = -\frac{b^2}{a^2 \lambda_1}.$$

Substituting this value of λ_2 in (1), we have

$$\lambda = \lambda_1, \quad \ldots \ldots \ldots \ldots \quad (2)$$

the slope of the original set of chords.

Therefore, *if one of two diameters bisects chords parallel to the other, the second will also bisect all chords parallel to the first.* Such diameters are called *conjugate.*

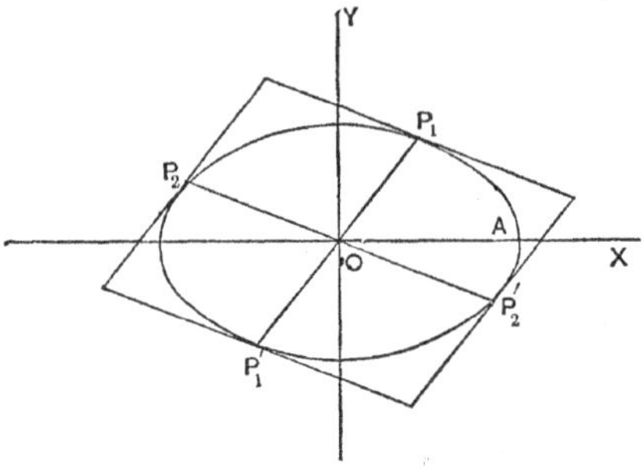

In the figure, $P_1' P_1$ and $P_2' P_2$ represent a pair of conjugate diameters of the ellipse, and each of these lines will bisect all chords parallel to the other.

The product of the slopes of a pair of conjugate diameters is constant; for we readily obtain from equations (1) and (2)

$$\lambda_1 \lambda_2 = -\frac{b^2}{a^2} \quad \ldots \ldots \ldots \ldots \quad (3)$$

NOTE. The results of this section may be obtained at once from equation (7) of the last section by writing it in the form

$$\lambda_1 = -\frac{b^2}{a^2 \lambda},$$

where λ_1 bears the same relation to λ which, in the original form, λ bore to λ_1. From this we may draw the inferences stated in this section.

Equation (3) shows that the product of the slopes of conjugate diameters is constantly negative; therefore one of these slopes must be positive and the other negative. It follows that the inclination of one of the diameters must be between 0° and 90°, and of the other between 90° and 180°, except in the special case where the diameters are the axes of the ellipse. (v. § 25, Rem.)

Therefore *conjugate diameters of an ellipse lie on opposite sides of the minor axis.*

94. Changing the sign of b^2 in the equations of the last two sections, we have for the equation of a diameter of the hyperbola bisecting chords with a slope λ_1.

$$y = \frac{b^2}{a^2 \lambda_1} x; \quad \ldots\ldots\ldots\ldots \quad (1)$$

and for the product of the slopes of conjugate diameters

$$\lambda_1 \lambda_2 = \frac{b^2}{a^2}. \quad \ldots\ldots\ldots\ldots \quad (2)$$

Let the student verify these equations by finding the equation of a diameter of an hyperbola, and show that if one diameter of an hyperbola bisects all chords parallel to the other, the second bisects all chords parallel to the first.

In the figure $P_1' P_1$ and $P_2' P_2$ represent a pair of conjugate diameters; $P_1' P_1$ extended bisects all chords of either branch of the hyperbola, which are parallel to $P_2' P_2$; and $P_2' P_2$ extended bisects all chords lying between the two branches of the curve and parallel to $P_1' P_1$. We limit $P_2' P_2$ by the conjugate hyperbola, as will be explained later.

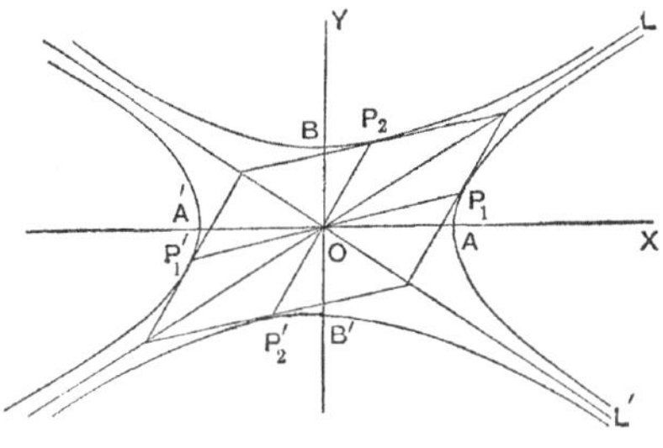

It follows from equation (2) that *conjugate diameters of an hyperbola lie on the same side of the conjugate axis*, as shown in the figure.

Also, the numerical value of the slope of one diameter must be less than $\frac{b}{a}$, and of the other greater than $\frac{b}{a}$, except in the case where each equals $\frac{b}{a}$ and the diameters coincide with one of the asymptotes.

We learn from this that one diameter has a less inclination than the asymptote, and the other a greater; therefore one lies in the angle $L' O L$, and the other in the supplementary angle. We may also state this fact as follows: —

Of two conjugate diameters of an hyperbola, only one meets the hyperbola, the other meeting the conjugate hyperbola.

EXAMPLES.

(1.) Find the equation of the diameter of the ellipse

$$\frac{x^2}{16} + \frac{y^2}{9} = 1,$$

which bisects the chord whose equation is

$$3x + y + 2 = 0.$$

Ans. $3x - 16y = 0$.

(2.) What is the equation of the diameter conjugate to the one found in the last example?

Ans. $3x + y = 0$.

(3.) Find the equations of a pair of conjugate diameters of the hyperbola

$$x^2 - 12y^2 = 48,$$

one of which bisects the chord whose equation is

$$2x - 9y = 3.$$

Ans. $3x = 8y$ and $2x = 9y$.

(4.) Find the equation of the diameter of the ellipse

$$4x^2 + 9y^2 = 144,$$

which has for one extremity the point $(3\sqrt{3}, 2)$; what is the other extremity of this diameter?

Ans. $2x = 3\sqrt{3}\,y$, $(-3\sqrt{3}, -2)$.

(5.) Find the equation and extremities of the diameter conjugate to the one found in the last example.

Ans. $2x + \sqrt{3}\,y = 0$, $(3, -2\sqrt{3})$ $(-3, 2\sqrt{3})$.

(6.) One extremity of the diameter of the ellipse

$$\frac{x^2}{a^2} + \frac{y^2}{b^2} = 1$$

is the point (x_1, y_1); what is the other extremity? what the equation of the diameter?

Ans. $(-x_1, -y_1)$, $y_1 x = x_1 y$.

(7.) What is the equation of the diameter conjugate to the one found in the last example?

Ans. $\frac{x_1 x}{a^2} + \frac{y_1 y}{b^2} = 0$.

(8.) Remembering that (x_1, y_1) is a point of the curve, prove that the extremities of the diameter of Ex. 7 are the points

$$\left(\frac{a}{b} y_1, -\frac{b}{a} x_1\right) \text{ and } \left(-\frac{a}{b} y_1, \frac{b}{a} x_1\right).$$

(9.) Show that the tangent to an ellipse at any point of the curve is parallel to the diameter conjugate to the one which passes through that point; and that tangents at opposite extremities of any diameter are parallel. Prove the same property for the hyperbola.

(10.) Find the equation of the diameter of the hyperbola

$$\frac{x^2}{a^2} - \frac{y^2}{b^2} = 1$$

which meets the curve in the point P_1; also, the equation of the conjugate diameter.

Ans. $y_1 x - x_1 y = 0$, $\frac{x_1 x}{a^2} - \frac{y_1 y}{b^2} = 0$.

(11.) Show that the coördinates of the points where the second diameter meets the hyperbola are imaginary, their values being

$$\left(\frac{a}{b}y_1\sqrt{-1}, \frac{b}{a}x_1\sqrt{-1}\right) \text{ and } \left(-\frac{a}{b}y_1\sqrt{-1}, -\frac{b}{a}x_1\sqrt{-1}\right).$$

(12.) Show that the second diameter meets the hyperbola conjugate to the one given in example (10) in the points

$$\left(\frac{a}{b}y_1, \frac{b}{a}x_1\right), \left(-\frac{a}{b}y_1, -\frac{b}{a}x_1\right).$$

(13.) Find the equation of the diameter which bisects all chords with a slope λ_1 in the hyperbola conjugate to

$$\frac{x^2}{a^2} - \frac{y^2}{b^2} = 1,$$

and show that it coincides with the corresponding diameter of this hyperbola.

(14.) Show that in the figure of § 94, $P_1' P_1$ and $P_2' P_2$ are a pair of conjugate diameters for each of the two conjugate hyperbolas.

(15.) Find the equation of the diameter of a parabola, which bisects all chords whose slope is λ_1; show that all diameters of a parabola are parallel to the axis.

Ans. $y = \dfrac{m}{\lambda_1}$.

(16.) Prove that when a parabola lies on the right of OY (i. e. when m is positive), the diameter bisecting parallel chords whose inclination is less than 90° lies above the axis of the curve; the diameter bisecting chords whose inclination is greater than 90° lies below the axis. State the corresponding theorem when m is negative. What is the diameter which bisects chords perpendicular to the axis?

95. By means of the results of example (8) of the last section we can find the lengths of a pair of conjugate diameters of an ellipse in terms of the coördinates of an extremity P_1 of one of them. For, since the centre is the middle point of each diameter, if we let a' and b' represent the halves of the conjugate diameters, we have

$$a'^2 = x_1^2 + y_1^2, \quad \ldots \ldots \quad (1)$$

and

$$b'^2 = \frac{a^2}{b^2} y_1^2 + \frac{b^2}{a^2} x_1^2 . \quad \ldots \ldots \quad (2)$$

Now P_1 is a point on the ellipse, and therefore

$$y_1^2 = \frac{b^2}{a^2}(a^2 - x_1^2) \quad \ldots \ldots \quad (3)$$

Substituting this value in equations (1) and (2), we have

$$a'^2 = x_1^2 + b^2 - \frac{b^2}{a^2} x_1^2$$

$$= b^2 + \frac{a^2 - b^2}{a^2} x_1^2$$

$$= b^2 + e^2 x_1^2, \quad \ldots \ldots \quad (4)$$

and

$$b'^2 = a^2 - x_1^2 + \frac{b^2}{a^2} x_1^2$$

$$= a^2 - e^2 x_1^2 . \quad \ldots \ldots \quad (5)$$

Equations (4) and (5) give the squares of the half-diameters in terms of x_1, y_1, and quantities which are known when the ellipse is given, and from them the lengths of the diameters may be found.

96. For the hyperbola, one of the semi-diameters is imaginary and the expression for its length contains the factor $\sqrt{-1}$, for we have shown in § 94, Ex. 11, that only one of a pair of conjugate diameters meets the hyperbola in real points. Using a' to represent the real semi-diameter, and b' to represent the real factor in the expression for the imaginary semi-diameter, — a use analogous to that of b (v. § 69),—we may write for the hyperbola

$$a'^2 = e^2 x_1^2 - b^2, \quad \ldots \ldots \ldots \ldots \quad (1)$$

and

$$b'^2 = e^2 x_1^2 - a^2. \quad \ldots \ldots \ldots \ldots \quad (2)$$

EXAMPLES.

(1.) Prove that the sum of the squares of any two conjugate semi-diameters of an ellipse is constant and equal to the sum of the squares on the semi-axes.

(2.) State and prove the theorem for the hyperbola corresponding to that given in the last example for the ellipse.

(3.) Prove that conjugate diameters of an equilateral hyperbola are equal.

(4.) Show that the ellipse has a pair of equal conjugate diameters, and show that they coincide with the asymptotes of an hyperbola which has the same axes as the ellipse.

(5.) Prove that the equal conjugate diameters of an ellipse bisect the lines joining the extremities of the axes.

(6.) Show that the perpendicular from the centre of an ellipse upon a tangent at a point P_1 of the curve is equal to $\frac{a\,b}{b'}$, where b' is the semi-diameter conjugate to the one which passes through P_1.

(7.) Prove that all parallelograms circumscribing a given ellipse, and having their sides parallel to a pair of conjugate diameters, have the same area.

(8.) Find the equations of the chords of an ellipse, which connect the extremities of a pair of conjugate diameters, one of which meets the curve in P_1.

Ans. $b\,(b\,x_1 - a\,y_1)\,x + a\,(a\,y_1 + b\,x_1)\,y - a^2\,b^2 = 0$

is the equation of one of the chords.

(9.) Find the locus of the middle point of the chord of the last example as the point P_1 moves around the ellipse.

97. Two points C and D are said to divide the straight line AB *harmonically* if one of them divides this line internally in the same ratio as the other divides it externally; *i.e.*, having regard to sign, if

$$\frac{CA}{CB} = -\frac{DA}{DB} \quad \ldots\ldots\ldots\ldots \quad (1)$$

This equation may also be written

$$\frac{AC}{AD} = -\frac{BC}{BD} \quad \ldots\ldots\ldots\ldots \quad (2)$$

The left-hand side of (2) is the ratio in which the point A divides the line CD, while the right-hand side is the negative of the ratio in which B divides this line. Accordingly (2) expresses the fact that A and B divide the line CD harmonically. Hence the theorem:

If the points C, D divide the line A B harmonically, then conversely the points A, B divide the line C D harmonically.

98. If the three points A, B, C are given, we can in general construct a fourth point D so that C, D divide AB harmonically. This is evident from Plane Geometry, since if C divides AB internally all we have to do is to construct the point dividing this line externally in the same ratio, and vice versa. Analytically we can see the same fact as follows. Let the coördinates of A be (x_1, y_1), of B, (x_2, y_2), and let $m_1 : m_2$ be the ratio in which C divides AB. Then (v. p. 9) the coördinates of C are

$$\left(\frac{m_2 x_1 - m_1 x_2}{m_2 - m_1}, \quad \frac{m_2 y_1 - m_1 y_2}{m_2 - m_1}\right) \quad \ldots\ldots \quad (1)$$

and the point D which divides $A\,B$ in the ratio $-m_1 : m_2$ has the coördinates

$$\left(\frac{m_2 x_1 + m_1 x_2}{m_2 + m_1}, \quad \frac{m_2 y_1 + m_1 y_2}{m_2 + m_1}\right) \quad \ldots\ldots \quad (2)$$

We thus see that there is no difficulty in finding the point D except in the one case in which $m_1 = -m_2$; *i.e.*, when C bisects the line $A\,B$. In this case it is also obvious geometrically that the point D cannot be constructed, since every point on the line $A\,B$ produced is nearer to one end than to the other.

It is often more convenient to use a single letter to denote the ratio in which C divides $A\,B$. If we write

$$\mu = \frac{m_1}{m_2}$$

the formulæ (1) and (2) for the coördinates of C and D take the form

$$\left(\frac{x_1 - \mu x_2}{1 - \mu}, \quad \frac{y_1 - \mu y_2}{1 - \mu}\right) \quad \ldots\ldots\ldots\ldots \quad (3)$$

$$\left(\frac{x_1 + \mu x_2}{1 + \mu}, \quad \frac{y_1 + \mu y_2}{1 + \mu}\right) \quad \ldots\ldots\ldots \quad (4)$$

99. By the *polar* of a fixed point P with regard to a given conic is meant the locus determined as follows. Through P a straight line is passed cutting the conic in C and D. On this line a fourth point Q is constructed so that P, Q divide $C\,D$ harmonically. The line $P\,C\,D$ is then allowed to revolve about P, the point Q being constructed on every position of the line in the manner just indicated. The locus of Q, which we shall prove to be in all cases a straight line, is called the polar of P.

Let us consider first the case in which the conic is an ellipse.

As usual we will use the major and minor axes of this ellipse as coördinate axes, writing its equation in the form

$$b^2 x^2 + a^2 y^2 = a^2 b^2 \ldots \ldots \ldots \ldots (1)$$

The coördinates of P shall be (x_1, y_1). Call the coördinates of the moving point Q, whose locus we want to find (x', y'). Since by hypothesis P, Q divide CD harmonically, conversely

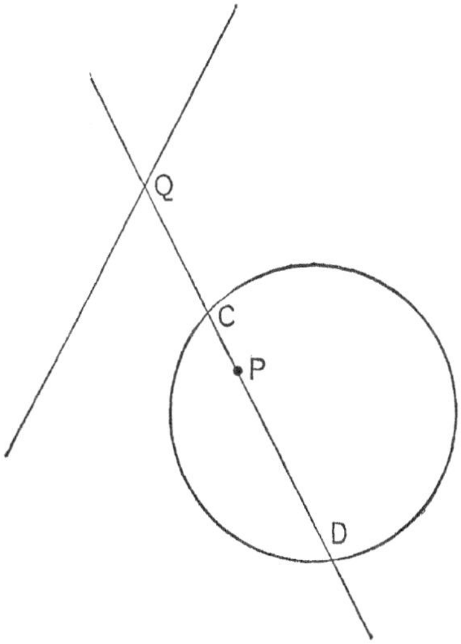

C, D must divide PQ harmonically (v. § 97). We will introduce as an auxiliary variable (v. § 56) the ratio in which the point C divides the line PQ. Calling this ratio μ, we have by formulæ (3) and (4) of § 98 as the coördinates of C and D respectively

$$\left(\frac{x_1 - \mu x'}{1 - \mu}, \frac{y_1 - \mu y'}{1 - \mu}\right) \ldots \ldots \ldots (2)$$

$$\left(\frac{x_1 + \mu x'}{1 + \mu}, \frac{y_1 + \mu y'}{1 + \mu}\right) \ldots \ldots \ldots (3)$$

Since C and D lie on the ellipse, we have the following two equations obtained by substituting the coördinates (2) and (3) in (1) and then clearing of fractions:

$$b^2(x_1 - \mu x')^2 + a^2(y_1 - \mu y')^2 = a^2 b^2 (1 - \mu)^2 \quad . \; . \quad (4)$$

$$b^2(x_1 + \mu x')^2 + a^2(y_1 + \mu y')^2 = a^2 b^2 (1 + \mu)^2 \quad . \; . \quad (5)$$

Between these two equations we must eliminate μ. This can be done by subtracting (4) from (5), getting

$$2\mu b^2 x_1 x' + 2\mu a^2 y_1 y' = 2\mu a^2 b^2 \quad . \; . \; . \; . \; . \quad (6)$$

The factor 2μ can be cancelled out from this equation, and we then have as the equation of the polar, after dropping the accents,

$$b^2 x_1 x + a^2 y_1 y = a^2 b^2 \quad . \; . \; . \; . \; . \; . \; . \; . \; . \quad (7)$$

or

$$\frac{x_1 x}{a^2} + \frac{y_1 y}{b^2} = 1 \quad . \; . \; . \; . \; . \; . \; . \; . \; . \; . \; . \; . \; . \quad (8)$$

This equation, being of the first degree, represents a straight line.

Replacing b^2 by $-b^2$, we find as the polar of the point (x_1, y_1) with regard to the hyperbola

$$\frac{x^2}{a^2} - \frac{y^2}{b^2} = 1, \quad . \; . \; . \; . \; . \; . \; . \; . \; . \; . \quad (9)$$

the straight line

$$\frac{x_1 x}{a^2} - \frac{y_1 y}{b^2} = 1 \quad . \; . \; . \; . \; . \; . \; . \; . \; . \; . \quad (10)$$

Show by the same method that the polar of the point (x_1, y_1) with regard to the parabola

$$y^2 = 2mx \quad . \; . \; . \; . \; . \; . \; . \; . \; . \; . \; . \; . \; . \; . \quad (11)$$

is the straight line

$$y_1 y = m(x + x_1) \quad . \; . \; . \; . \; . \; . \; . \; . \; . \; . \quad (12)$$

If $a = b$, the ellipse (1) becomes the circle

$$x^2 + y^2 = a^2 \quad \dots\dots\dots\dots \quad (13)$$

Accordingly the polar of (x_1, y_1) with regard to this circle is the straight line

$$x_1 x + y_1 y = a^2 \quad \dots\dots\dots\dots \quad (14)$$

EXERCISES.

(1.) Find the polar of the point (2, − 1) with regard

(*a*) to the ellipse $\frac{x^2}{9} + \frac{y^2}{4} = 1$;
(*b*) to the hyperbola $2x^2 - 3y^2 = 1$;
(*c*) to the parabola $y^2 = 6x$.

(2.) Prove that the polar of a point with regard to a circle is perpendicular to the line joining the point to the centre of the circle; and that the radius of the circle is a mean proportional between the distances of the point and its polar from the centre of the circle.

(3.) Find the equation of the polar of the point (x_1, y_1) with regard to the hyperbola conjugate to (9).

Hence prove that the polars of a point with regard to two conjugate hyperbolas are parallel. What more can you say about the relative position of these two lines?

(4.) Find the equation of the polar of the point (x_1, y_1) with regard to the conic (2) of § 89.

Hence show that the polar of a focus of a conic is the corresponding directrix.

100. It will be noticed that formulæ (8), (10), (12), and (14) for the polars of (x_1, y_1) with regard to the ellipse, hyperbola, parabola, and circle are identical with the formulæ we obtained in the last chapter (§§ 74, 75, 76) and in § 49 for the tangents to these curves at the point (x_1, y_1). The difference is that in finding the tangent we assumed that (x_1, y_1) lies on the curve, while we have now assumed that (x_1, y_1) does not lie on the curve. In fact the definition of the polar which we have given obviously breaks down if P lies on the conic. It is customary, however, in this case to speak of the tangent of the conic at P as being the polar of P. In this way the formulæ of the last section represent the polar of (x_1, y_1) even if this point lies on the conic.

There is one other case in which our definition of a polar breaks down, namely when P lies at the centre of the conic; for in this case all chords through P are bisected there, so that the point Q can never be constructed. *The centre of a conic has no polar.*

101. If P lies on the convex side of the conic, we can draw two tangents from it touching the conic at P_1 and P_2 (v. the

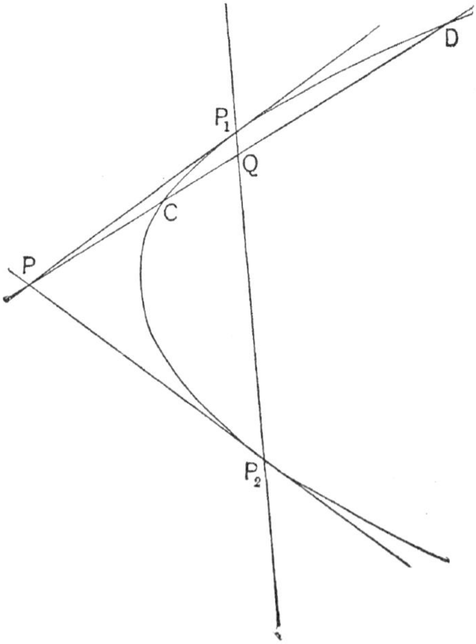

figure). *In this case the line P_1, P_2 is the polar of P.* For if we allow the secant $P\,C\,D$ to turn around P and approach the position of the tangent $P\,P_1$ as a limit, both the points C and D approach the point P_1 as their limits, and since Q always lies between C and D, it must also approach P_1 as its limit. Therefore the locus of Q points directly towards P_1 and, if continued, passes through P_1. In the same way we see that the polar passes through P_2; so that, being a straight line, it must coincide with the line $P_1\,P_2$.

102. We have seen that every point (with the single exception of the centre of the conic) has a definite polar. If the line $A\,B$ is the polar of P, then conversely P is called the *pole* of $A\,B$; and now it is a fact that *every line (with the exception of lines through the centre of the conic) has a definite pole.*

In order to prove this theorem let us suppose first that our conic is the ellipse

$$\frac{x^2}{a^2} + \frac{y^2}{b^2} = 1. \quad \ldots \ldots \ldots \ldots \quad (1)$$

Let the given line be

$$A\,x + B\,y + C = 0. \quad \ldots \ldots \ldots \quad (2)$$

In order that (x_1, y_1) should be the pole of (2) it is necessary and sufficient that its polar

$$\frac{x_1\,x}{a^2} + \frac{y_1\,y}{b^2} - 1 = 0 \quad \ldots \ldots \ldots \quad (3)$$

should coincide with (2), and the condition for this (v. § 30) is that the coefficients of (2) and (3) be proportional:

$$\frac{x_1}{a^2} = -\frac{A}{C}, \quad \frac{y_1}{b^2} = -\frac{B}{C}.$$

Accordingly

$$x_1 = -\frac{A\,a^2}{C}, \quad y_1 = -\frac{B\,b^2}{C}$$

is the pole of (2),—a perfectly definite point if C is not zero, that is, if (2) does not pass through the centre of the ellipse.

A similar proof holds in the cases of the hyperbola and the parabola.

103. The most important property of poles and polars is expressed in the following:

THEOREM.—*If two points P_1 and P_2 and their polars A_1B_1 and A_2B_2 are given, and if A_1B_1 passes through P_2, then A_2B_2 will pass through P_1.*

We will prove this theorem for the case of the ellipse, the proofs for the hyperbola or the parabola being almost identical.

Let the coördinates of P_1 and P_2 be (x_1, y_1) and (x_2, y_2) respectively. Then the equations of $A_1 B_1$ and $A_2 B_2$ are

$$\frac{x_1 x}{a^2} + \frac{y_1 y}{b^2} = 1, \quad \ldots \ldots \ldots \quad (1)$$

$$\frac{x_2 x}{a^2} + \frac{y_2 y}{b^2} = 1 \quad \ldots \ldots \ldots \quad (2)$$

Since by hypothesis (x_2, y_2) lies on (1) we have

$$\frac{x_1 x_2}{a^2} + \frac{y_1 y_2}{b^2} = 1;$$

but this is precisely the condition that the point (x_1, y_1) should lie on (2). This proves the theorem.

From this theorem it follows at once that *if several points lie on a line, their polars intersect on the pole of that line, and if several lines meet in a point, their poles lie on the polar of this point.*

EXAMPLES.

1. Find the pole of the line $x + y = 1$ with regard
   (*a*) to the ellipse $x^2 + 4y^2 = 1$;
   (*b*) to the parabola $y^2 = 4x$.

2. Prove that the tangents at the extremity of any focal chord of a conic intersect on the directrix (v. Ex. 4, § 99).

3. If F is a focus of a conic, and P is a point on the corresponding directrix, prove that the line PF is perpendicular to the polar of P.

4. Show that the fact established in § 101 is a special case of the theorem of § 103.

CHAPTER IX.

THE GENERAL EQUATION OF THE SECOND DEGREE.

104. We saw in Chapter III that a straight line is a curve of the first degree, and, conversely, that every curve of the first degree is a straight line.

In Chapter VII we have seen that the three conic sections are curves of the second degree. These, however, are not the only curves of this kind, since two straight lines may be represented by a single equation of the second degree (v. § 59). It is our purpose in this chapter to show that every equation of the second degree, if it has a real locus, represents either one of the conic sections or two straight lines.

The general equation of the second degree is

$$Ax^2 + Bxy + Cy^2 + Dx + Ey + F = 0. \quad . \quad . \quad . \quad (1)$$

We begin by examining the special case of this equation in which $B = 0$:

$$Ax^2 + Cy^2 + Dx + Ey + F = 0. \quad . \quad . \quad . \quad . \quad . \quad (2)$$

105. Before considering equation (2) of the last section in all its generality, let us look at two special cases.

Suppose first that $D = E = 0$ while neither A nor C is zero. The equation becomes

$$Ax^2 + Cy^2 + F = 0, \quad . \quad . \quad . \quad . \quad . \quad . \quad . \quad . \quad . \quad (1)$$

or, as we may write it,

$$-\frac{A}{F}x^2 - \frac{C}{F}y^2 = 1, \quad \ldots\ldots\ldots\ldots \quad (2)$$

or, finally,

$$\frac{x^2}{-\frac{F}{A}} + \frac{y^2}{-\frac{F}{C}} = 1 \quad \ldots\ldots\ldots\ldots \quad (3)$$

If these denominators are positive we may denote their square roots by a and b, and (3) takes the standard form for the equation of an ellipse (v. § 63, (7)). If the first denominator is positive and the second negative we have in the same way the standard form for the equation of the hyperbola. If the first denominator is negative and the second positive equation (3) takes the form

$$\frac{x^2}{a^2} - \frac{y^2}{b^2} = -1,$$

which, as we know (v. § 86), also represents a hyperbola, but one whose transverse axis lies on the axis of Y. Finally, if the two denominators in (3) are negative, the equation (3) has no locus since its right-hand side, being essentially negative, cannot be equal to 1. In this case (3) is often spoken of as *an imaginary ellipse.*

In this discussion we have tacitly assumed that F is not zero, since in passing from (1) to (2) we divided by F. If $F = 0$ equation (1) becomes

$$Ax^2 + Cy^2 = 0 \quad \ldots\ldots\ldots\ldots \quad (4)$$

If A and C have the same sign, this equation is obviously satisfied only by the values $x = 0, y = 0$. The locus of (4) is therefore a single point, the origin; and (4) is spoken of as a *null ellipse* (cf. the use of the term *null circle* in § 46). If A

and C have different signs we may suppose A positive and C negative (as otherwise we could change the signs throughout in (4)), and then (4) may be written

$$a^2 x^2 - b^2 y^2 = 0,$$

an equation which represents the two straight lines

$$ax - by = 0,$$

$$ax + by = 0,$$

which intersect at the origin.

Thus we have seen that, it being assumed that neither A nor C is zero, equation (1) represents an ellipse (real, null, or imaginary), an hyperbola, or two intersecting straight lines.

106. As a second special case of equation (2), § 104, let us suppose that $A = E = 0$, while C is not zero:

$$Cy^2 + Dx + F = 0 \quad . \quad . \quad . \quad . \quad . \quad . \quad . \quad . \quad . \quad . \quad (1)$$

If $D = 0$ this equation may be written

$$y = \pm\sqrt{-\frac{F}{C}} \quad . \quad . \quad . \quad . \quad . \quad . \quad . \quad . \quad . \quad . \quad . \quad . \quad (2)$$

This equation has no real locus if C and F have the same sign. It represents two lines parallel to the axis of x if C and F have opposite signs; and finally if $F = 0$ we have no longer two distinct lines, but only the single line $y = 0$. These three cases may be conveniently characterized by saying we have two parallel lines, real and distinct, real and coincident, or imaginary.

If, however, D is not zero, we write (1) in the form

$$Cy^2 = -D\left(x + \frac{F}{D}\right), \quad . \quad . \quad . \quad . \quad . \quad . \quad . \quad . \quad . \quad (3)$$

and now we make use of a device, which we shall have to employ again presently. We make the transformation of coördinates

$$x' = x + \frac{F}{D},$$

$$y' = y,$$

that is (v. § 41 (3)) we shift our coördinate axes, without turning them, so that the origin of the new system lies at the point $\left(-\frac{F}{D}, 0\right)$. After this transformation of coördinates the equation of the curve (3) has the form

$$y'^2 = -\frac{D}{C}x' \quad \ldots\ldots\ldots\ldots \quad (4)$$

and this is in the standard form, $y^2 = 2\,m\,x$, for the equation of a parabola. Accordingly

Equation (1) *represents either a parabola or two parallel straight lines* (*distinct, coincident, or imaginary*).

Precisely the same reasoning applies to the equation

$$Ax^2 + Ey + F = 0 \quad \ldots\ldots\ldots\ldots \quad (5)$$

(where we assume that A is not zero) except that the x and y coördinates are interchanged.

107. We are now in a position to take up the consideration of equation (2), § 104. We will first illustrate the method to be used by a numerical example.

Suppose we have the equation

$$2x^2 - 3y^2 + 4x + 5y + 3 = 0.$$

Let us rewrite this equation, taking first the two terms in x, then the two terms in y, and last the constant term. We will also take out the factor 2 (the coefficient of x^2) before the two terms in x, and the factor -3 (the coefficient of y^2) before the two terms in y:

$$2(x^2 + 2x) - 3(y^2 - \tfrac{5}{3}y) + 3 = 0.$$

Let us now complete the squares for the two expressions in parentheses by adding 2 to the first term and $-\frac{25}{12}$ to the second.

$$2(x + 1)^2 - 3(y - \tfrac{5}{6})^2 + (3 - 2 + \tfrac{25}{12}) = 0.$$

If we now make the transformation of coördinates

$$x' = x + 1,$$
$$y' = y - \tfrac{5}{6},$$

i.e., if we shift the coördinate axes without turning so that the origin lies at the point $(-1, \frac{5}{6})$ the equation of the curve takes the form

$$2x'^2 - 3y'^2 + \tfrac{37}{12} = 0.$$

This equation is of the form considered in § 105. Following the method there indicated we throw it into the form

$$\frac{x'^2}{\frac{37}{24}} - \frac{y'^2}{\frac{37}{36}} = -1.$$

Accordingly our original equation represents an hyperbola whose centre is at the point $(-1, \frac{5}{6})$, whose transverse axis is parallel to the axis of Y and is of length $2\sqrt{\frac{37}{24}} = \frac{1}{6}\sqrt{222}$, and whose conjugate axis is of length $2\sqrt{\frac{37}{36}} = \frac{1}{3}\sqrt{37}$.

Precisely the method just explained applies to any equation of the form (2), § 104, provided neither A nor C is zero. We first write the equation in the form

$$A\left(x^2 + \frac{D}{A}x\right) + C\left(y^2 + \frac{E}{C}y\right) + F = 0,$$

and then, by completing the square in each of the parentheses, in the form

$$A\left(x+\frac{D}{2A}\right)^2+C\left(y+\frac{E}{2C}\right)^2+F-\frac{D^2}{4A}-\frac{E^2}{4C}=0.$$

The transformation of coördinates

$$x'=x+\frac{D}{2A},$$

$$y'=y+\frac{E}{2C},$$

reduces this equation to

$$Ax'^2+Cy'^2+\left(F-\frac{D^2}{4A}-\frac{E^2}{4C}\right)=0,$$

and this equation is of the form considered in § 105.

108. If either A or C is zero the method must be slightly modified. Suppose, for instance, $A=0$. Then we can write equation (2), § 104, in the form

$$C\left(y^2+\frac{E}{C}y\right)+Dx+F=0 \quad \ldots\ldots\ldots \quad (1)$$

Completing the square, this equation becomes

$$C\left(y+\frac{E}{2C}\right)^2+Dx+F-\frac{E^2}{4C}=0 \quad \ldots\ldots \quad (2)$$

After the transformation of coördinates

$$\left.\begin{aligned} x'&=x, \\ y'&=y+\frac{E}{2C}, \end{aligned}\right\} \quad \ldots\ldots\ldots\ldots\ldots \quad (3)$$

the equation of the curve reduces to

$$Cy'^2+Dx'+\left(F-\frac{E^2}{4C}\right)=0, \quad \ldots\ldots \quad (4)$$

and this equation is of the form (1), § 106.

Similarly if $C = 0$ the equation can be reduced to the form (5), § 106.

109. Thus we have proved that if neither A nor C is zero equation (2), § 104, represents an ellipse (real, null, or imaginary), an hyperbola, or two intersecting straight lines; while if A or C is zero it represents either a parabola or two parallel straight lines (distinct, coincident, or imaginary).

Moreover, the method we have used enables us in every case to determine not only what kind of curve is represented by the equation, but just how it is situated.

EXERCISES.

(1.) What curves are represented by the following equations? Draw diagrams to show how they are situated.

(*a.*) $2x^2 + 3y^2 - 5 = 0.$

(*b.*) $2x^2 - 3y^2 + 5 = 0.$

(*c.*) $y^2 - 2x - 3 = 0.$

(*d.*) $y^2 + 3x + 2 = 0.$

(*e.*) $x^2 - 3y + 5 = 0.$

(*f.*) $2x^2 + 5y^2 - 3x + 2y - 4 = 0.$

(*g.*) $9x^2 + 4y^2 - 18x + 24y + 27 = 0.$

(*h.*) $2x^2 + 3x - 5y + 2 = 0.$

(2.) Show that the condition that (2), § 104, represents two straight lines is

$$4ACF - CD^2 - AE^2 = 0.$$

110. We come now to the general equation of the second degree, (1), § 104, where B is not zero. Let us try to simplify this equation by turning the coördinate axes through the angle ϑ. The formulæ for this transformation (v. (7), § 42) are

$$\left.\begin{array}{l} x = x' \cos\vartheta - y' \sin\vartheta, \\ y = x' \sin\vartheta + y' \cos\vartheta \end{array}\right\} \quad \ldots\ldots\ldots \quad (1)$$

Equation (1), § 104, now takes the form

$$A' x'^2 + B' x' y' + C' y'^2 + D' x' + E' y' + F = 0, \quad (2)$$

where

$$A' = A \cos^2\vartheta + B \sin\vartheta \cos\vartheta + C \sin^2\vartheta, \quad \ldots\ldots\ldots \quad (3)$$

$$B' = -2A \sin\vartheta \cos\vartheta + B \cos^2\vartheta - B \sin^2\vartheta + 2C \sin\vartheta \cos\vartheta, \quad \ldots \quad (4)$$

$$C' = A \sin^2\vartheta - B \sin\vartheta \cos\vartheta + C \cos^2\vartheta, \quad \ldots\ldots\ldots \quad (5)$$

$$D' = D \cos\vartheta + E \sin\vartheta, \quad \ldots\ldots\ldots \quad (6)$$

$$E' = -D \sin\vartheta + E \cos\vartheta. \quad \ldots\ldots\ldots \quad (7)$$

The second of these coefficients may be written more simply as follows:

$$B' = B \cos 2\vartheta - (A - C) \sin 2\vartheta. \quad \ldots\ldots \quad (8)$$

Accordingly B' will be zero if

$$B \cos 2\vartheta = (A - C) \sin 2\vartheta,$$

i.e., if

$$\cot 2\vartheta = \frac{A - C}{B}. \quad \ldots\ldots\ldots \quad (9)$$

If we remember that as an angle increases from 0 to 180° its cotangent decreases from $+\infty$ to $-\infty$, we see that there is just one value of 2ϑ less than 180° which satisfies (9), and therefore just one *acute* value of ϑ.

If, then, we turn the coördinate axes through the acute angle ϑ determined by (9), equation (1), § 104, takes the form

$$A' x'^2 + C' y'^2 + D' x' + E' y' + F = 0,$$

This equation is of the form (2), § 104. Accordingly we have proved that every equation of the second degree which has a real locus represents either a conic section or two straight lines.

EXAMPLES.

(1.) Determine what curves are represented by the following equations and how they are situated. Illustrate your answers by diagrams:

(*a*) $x^2 + 4xy + y^2 + 5 = 0.$

(*b*) $x^2 + 2xy - y^2 + 1 = 0.$

(*c*) $2x^2 + 4xy + 2y^2 - 3x + y = 0.$

(*d*) $2x^2 - 3xy + 2y^2 - 3x + 2y - 5 = 0.$

This last equation, like those which precede it, can be treated by the method of § 110. Let us, however, shift the coördinate axes without turning them so that the origin lies at the point (a, b).

$$x = x' + a,$$

$$y = y' + b.$$

Equation (*d*) then takes the form

$$2x'^2 - 3x'y' + 2y'^2 + (4a - 3b - 3)x' - (3a - 4b - 2)y'$$
$$+ (2a^2 - 3ab + 2b^2 - 3a + 2b - 5) = 0.$$

Now a and b can be so chosen that the coefficient of x' and y' are zero:

$$a = \tfrac{6}{7}, \quad b = \tfrac{1}{7}.$$

Thus if we shift to the new origin $(\frac{6}{7}, \frac{1}{7})$, equation (*d*) takes the form

$$14x^2 - 21xy + 14y^2 - 43 = 0,$$

and this equation can be much more easily treated than equation (*d*) by the method of § 110.

(2.) Discuss by the method just explained the equation

$$3x^2 + 5xy + y^2 + 2x - 3y + 1 = 0.$$

MISCELLANEOUS EXAMPLES.

(1.) Prove that the foot of the perpendicular from a focus of an hyperbola to an asymptote is at distances a and b from the centre and focus respectively.

(2.) A circle has its centre at the vertex O of a parabola whose focus is F, and the diameter of the circle is $3\ \overline{OF}$; show that the common chord of the curves bisects OF.

(3.) Prove that the semi-minor axis of an ellipse is a mean proportional between the parts of a tangent at an extremity of the major axis, cut off by conjugate diameters produced.

(4.) Prove that the parts of a normal to an hyperbola cut off by the transverse and conjugate axes respectively, are in the ratio $b^2 : a^2$.

(5.) Prove that if the major axis of an ellipse is equal to twice the minor axis, a straight line, equal to half the major axis, and which moves with one end on the upper half of the curve and the other on the lower half of the minor axis, is bisected by the major axis.

(6.) A line of constant length moves with its extremities on two straight lines at right angles to each other: show that the locus of a fixed point on the moving line is an ellipse with the segments of the line for semi-axes.

(7.) Adapt the last example to the case in which the fixed point is taken on the extension of the moving line.

(8.) Two equal rulers, AB, BC, are connected by a pivot at B; the extremity A is fixed, while C moves on a given straight line: find the locus of any fixed point of BC, taking the origin at A and AC for the axis of x.

(9.) Normals are drawn to an ellipse and the circumscribed circle at *corresponding* points; find the locus of their intersection.

Ans. A circle whose radius is $a + b$.

(10.) The ordinate of a point on an hyperbola is extended until its length equals a focal radius of the point; find the locus of its extremity.

(11.) Any straight line through the origin meets a parallel to OX drawn through a fixed point, B, on the axis of y, in P'; on OP', P is taken so that its ordinate equals BP'; find the locus of P.

Ans. A parabola.

(12.) Find the locus of the centre of a circle which passes through a fixed point and touches a given line. (Axes as in § 55, Ex. 13.)

Ans. A parabola.

(13.) Given the base and the product of the tangents of the base angles of a triangle; find the locus of the vertex.

SUGGESTION. Taking the axes as in § 55, Ex. 5, the tangents of the base angles are $\frac{y}{c+x}$ and $\frac{y}{c-x}$.

Ans. Letting m represent the given product, the locus is an ellipse or hyperbola, according as m is positive or negative.

(14.) Given the base and the product of the tangents of the halves of the base angles; find the locus of the vertex.

If we express the tangents of the half angles in terms of the sides, we shall find that the sum of the sides, $2s$ is given; the locus must therefore be an ellipse with the extremities of the base for its foci.

(15.) Show by means of the results of the last two examples that when the base and the sum of the sides of a triangle are given, the centre of the inscribed circle moves on an ellipse whose major axis is the base of the triangle.

(16.) Show that the equation of the ellipse referred to the major axis as the axis of x, and its left-hand extremity as origin, may be written in the form

$$y^2 = \frac{b^2}{a^2}(2\,a\,x - x^2).$$

(17.) By means of this equation, prove that the locus of the middle points of chords of an ellipse, which pass through the left-hand extremity of the major axis, is another ellipse. What are its axes, and how is it situated?

(18.) Prove that if the right-hand focus of an ellipse is the origin (the major axis being the axis of x), the radius connecting that focus with any point of the ellipse equals $a\,(1 - e^2) - e\,x$; and when the left-hand focus is the origin, the radius joining that focus with a point of the ellipse is $a\,(1 - e^2) + e\,x$.

(19.) By means of the results of the last example, obtain the polar equations of the ellipse referred to the right-hand and left-hand focus, respectively.

$$Ans.\ r = \frac{a\,(1 - e^2)}{1 + e\cos\phi},\quad r = \frac{a\,(1 - e^2)}{1 - e\cos\phi}.$$

(20.) Show that for the right-hand branch of the hyperbola the focal radii, when referred to the respective foci (v. Ex. 18), are

$$e\,x - a\,(1 - e^2) \text{ and } e\,x + a\,(1 - e^2);$$

changing the signs of these expressions, we have the focal radii for the left-hand branch.

(21.) From the results of the last example, show that the polar equations of the right-hand branch of an hyperbola referred to the foci are

$$r = -\frac{a\,(1 - e^2)}{1 - e\cos\phi} \text{ and } r = \frac{a\,(1 - e^2)}{1 - e\cos\phi}.$$

The negative values of r derived from these equations give points on the left-hand branch, or we may use for that branch the following equations, similarly obtained:—

$$r = \frac{a\,(1 - e^2)}{1 + e\cos\phi} \quad \text{and} \quad r = -\frac{a\,(1 - e^2)}{1 + e\cos\phi}.$$

(22.) Obtain the polar equation of the parabola referred to the focus, by transformation of coördinates.

Ans. $r = \dfrac{m}{1 - \cos\phi}.$

(23.) Find the polar equation mentioned in the last example from the definition of the curve, without assuming any other form of the equation of the parabola.

(24.) Discuss the polar equation of the parabola referred to the focus.

(25.) From the equation of the parabola referred to axis and directrix, obtain the polar equation when the intersection of these lines is the pole.

(26.) Find the locus of the middle points of chords of a parabola, which pass through (1) the vertex, (2) the focus, (3) the intersection of axis and directrix.

Ans. Parabolas whose parameters are one half that of the original curve.

(27.) Prove that if a straight line is drawn from the intersection of the directrix and axis of a parabola cutting the curve, the rectangle of the intercepts made by the curve is equal to the rectangle of the parts into which the parallel focal chord is divided by the focus.

(28.) Two parabolas have the same focus and the same transverse axis. A tangent to the first curve and a tangent to the second move so as to remain at right angles to each other. Find the locus of their intersection.

SUGGESTION. Take the common focus as origin.

Ans. The straight line parallel to the two directrices and half way between them.

(29.) Show, by the method of § 33, that the tangent to an ellipse or hyperbola at any point makes equal angles with the focal radii drawn to that point. (v. §§ 79, 81.)

(30.) Prove, as in the last example, that the tangent to a parabola at any point makes equal angles with the axis and the focal radius of that point.

(31.) Prove that tangents to the parabola at P_1 and P_2 intersect at the point whose coördinates are

$$y' = \frac{y_1 + y_2}{2}, \quad x' = \frac{y_1 y_2}{2m}.$$

(32.) From the results of the last example, show that the area of a triangle formed by three tangents to a parabola is one half the area of the triangle whose vertices are the points of tangency.

(33.) Prove that the triangle formed by a tangent to an hyperbola and the asymptotes has a constant area.

(34.) If from any point P' of a line $M' P'$, perpendicular to the axis $C O M'$ of a parabola whose vertex is O, a parallel to the axis is drawn meeting the curve in P''; show that, if $C O$ is made equal to $O M'$, the locus of the intersection of $O P'$ and $C P''$ is the original curve.

(35.) $P' P''$ is any chord of an ellipse at right angles to the major axis $A' A$; find the locus of the intersection of $A' P'$ and $A P''$.

Ans. An hyperbola with the same axes as the ellipse.

(36.) The right-hand extremity of a diameter of a circle is joined with the middle point of a parallel chord; find the locus of the intersection of this line with the radius drawn to the right-hand extremity of the chord.

Ans. A parabola.

(37.) Prove that the perpendicular from the centre of an ellipse upon a straight line which joins the ends of perpendicular diameters is of constant length. What is the locus of its extremity?

(38.) From a point P' of an ellipse straight lines are drawn to A' and A, the extremities of the major axis, and from A and A' perpendiculars are erected to $A' P'$ and $A P'$; show that the locus of their intersection is another ellipse, and find its axes.

(39.) Find the locus of the intersection of the ordinate of any point of an ellipse, produced, with the perpendicular from the centre to the tangent at that point.

(40.) A perpendicular is drawn from a focus of an ellipse to any diameter; find the locus of its intersection with the conjugate diameter.

Ans. A straight line perpendicular to the major axis.

(41.) Two perpendiculars are drawn from the extremities of a pair of conjugate diameters of an ellipse to the diameter whose equation is

$$y = x \tan \alpha;$$

show that the sum of the squares of the perpendiculars is

$$a^2 \sin^2 \alpha + b^2 \cos^2 \alpha.$$

(42.) Prove that the tangents to conjugate hyperbolas at the extremities of conjugate diameters intersect on the asymptotes.

(43.) Show that the locus of the intersection of tangents to an ellipse at the extremities of conjugate diameters is an ellipse; find its axes.

(44.) Show that the equation

$$xy = c$$

represents a rectangular hyperbola whose asymptotes are the coördinate axes.

(45.) Obtain by the method of § 49 the equation of the tangent to the rectangular hyperbola $xy = c$ at the point (x_1, y_1). *Ans.* $y_1 x + x_1 y = 2c$.

(46.) Obtain by the method of § 99 the equation of the polar of the point (x_1, y_1) with regard to the hyperbola $xy = c$.

(47.) Prove that if two rectangular hyperbolas are so situated that the axes of one are the asymptotes of the other, the polars of any point with regard to the two hyperbolas are perpendicular to each other. Hence show that these hyperbolas intersect at right angles.

(48.) Two concentric ellipses have axes lying on the same lines. Tangents are drawn to one ellipse. Find the locus of their poles with regard to the other.

(49.) Prove that the circle described on any focal radius of an ellipse as diameter is tangent to the circle described on the transverse axis of the ellipse as diameter.

(50.) State and prove the proposition analogous to (49) in the case of the parabola. Of the hyperbola.

(51.) A set of lines terminated by two rectangular axes pass through a fixed point. Find the equation of the locus of their middle points, and determine what this equation represents. *Ans.* A rectangular hyperbola.

(52.) In an equilateral hyperbola show that focal chords parallel to conjugate diameters are equal.

(53.) Parallel tangents are drawn to a set of confocal ellipses. Find the equation of the locus of the points of contact, and determine what this equation represents.

Ans. A rectangular hyperbola.

(54.) Find by the method of § 51 the equation of the diameter of the hyperbola $xy = c$ which bisects the chords whose slope is λ. Hence show that the angles between any pair of conjugate diameters of a rectangular hyperbola are bisected by the asymptotes.

(55.) Find by the method of § 49 the equation of the tangent to the conic

$$Ax^2 + Bxy + Cy^2 + Dx + Ey + F = 0$$

at the point $(x_1 y_1)$.

Ans. $$Ax_1 x + \frac{B}{2}(y_1 x + x_1 y) + Cy_1 y + \frac{D}{2}(x + x_1) + \frac{E}{2}(y + y_1) + F = 0.$$

(56.) Find the polar of (x_1, y_1) with regard to the conic of Ex. (55).

(57.) Show that the condition that the conic of Ex. (55) be a rectangular hyperbola is $A = -C$.

SHORT-TITLE CATALOGUE

OF THE

PUBLICATIONS

OF

JOHN WILEY & SONS,

NEW YORK.

LONDON: CHAPMAN & HALL, LIMITED.

ARRANGED UNDER SUBJECTS.

Descriptive circulars sent on application.
Books marked with an asterisk are sold at *net* prices **only.**
All books are bound in cloth unless otherwise stated.

AGRICULTURE.

Armsby's Manual of Cattle-feeding....................12mo, $1 75
Budd and Hansen's American Horticultural Manual:
  Part I.—Propagation, Culture, and Improvement....12mo, 1 50
  Part II.—Systematic Pomology. (*In preparation.*)
Downing's Fruits and Fruit-trees of America..............8vo, 5 00
Grotenfelt's Principles of Modern Dairy Practice. (Woll.)..12mo, 2 00
Kemp's Landscape Gardening..........................12mo, 2 50
Maynard's Landscape Gardening as Applied to Home Decoration. 12mo, 1 50
Sanderson's Insects Injurious to Staple Crops............12mo, 1 50
"   Insects Injurious to Garden Crops. (*In preparation.*)
"   Insects Injuring Fruits. (*In preparation.*)
Stockbridge's Rocks and Soils...........................8vo, 2 50
Woll's Handbook for Farmers and Dairymen.............16mo, 1 50

ARCHITECTURE.

Baldwin's Steam Heating for Buildings................12mo, 2 50
Berg's Buildings and Structures of American Railroads....4to, 5 00
Birkmire's Planning and Construction of American Theatres.8vo, 3 00
"   Architectural Iron and Steel..................8vo, 3 50
"   Compound Riveted Girders as Applied in Buildings. 8vo, 2 00
"   Planning and Construction of High Office Buildings. 8vo, 3 50
"   Skeleton Construction in Buildings..............8vo, 3 00
Briggs's Modern American School Buildings............8vo, 4 00
Carpenter's Heating and Ventilating of Buildings.........8vo, 4 00
Freitag's Architectural Engineering. 2d Edition, Rewritten.8vo, 3 50
"   Fireproofing of Steel Buildings................8vo, 2 50
Gerhard's Guide to Sanitary House-inspection............16mo, 1 00
"   Theatre Fires and Panics....................12mo, 1 50
Hatfield's American House Carpenter..................8vo, 5 00
Holly's Carpenters' and Joiners' Handbook.............18mo, 75
Kidder's Architect's and Builder's Pocket-book..16mo, morocco, 4 00
Merrill's Stones for Building and Decoration.............8vo, 5 00
Monckton's Stair-building...............................4to, 4 00

Patton's Practical Treatise on Foundations................8vo, 5 00
Siebert and Biggin's Modern Stone-cutting and Masonry..8vo, 1 50
Snow's Principal Species of Wood: Their Characteristic Properties. (*In preparation.*)
Wait's Engineering and Architectural Jurisprudence......8vo, 6 00
Sheep, 6 50
" Law of Operations Preliminary to Construction in Engineering and Architecture.....................8vo, 5 00
Sheep, 5 50
" Law of Contracts.................................8vo, 3 00
Woodbury's Fire Protection of Mills......................8vo, 2 50
Worcester and Atkinson's Small Hospitals, Establishment and Maintenance, and Suggestions for Hospital Architecture, with Plans for a Small Hospital..................12mo, 1 25
The World's Columbian Exposition of 1893..........Large 4to, 1 00

ARMY AND NAVY.

Bernadou's Smokeless Powder, Nitro-cellulose, and the Theory of the Cellulose Molecule.......................12mo, 2 50
* Bruff's Text-book Ordnance and Gunnery................8vo, 6 00
Chase's Screw Propellers and Marine Propulsion..........8vo, 3 00
Craig's Azimuth...4to, 3 50
Crehore and Squire's Polarizing Photo-chronograph......8vo, 3 00
Cronkhite's Gunnery for Non-commissioned Officers..24mo, mor., 2 00
* Davis's Elements of Law...............................8vo, 2 50
* " Treatise on the Military Law of United States..8vo, 7 00
* Sheep, 7 50
De Brack's Cavalry Outpost Duties. (Carr.)....24mo, morocco, 2 00
Dietz's Soldier's First Aid Handbook...........16mo, morocco, 1 25
* Dredge's Modern French Artillery..........4to, half morocco, 15 00
Durand's Resistance and Populsion of Ships................8vo, 5 00
* Dyer's Handbook of Light Artillery.................12mo, 3 00
Eissler's Modern High Explosives.........................8vo, 4 00
* Fiebeger's Text-book on Field Fortification........Small 8vo, 2 00
Hamilton's The Gunner's Catechism...................18mo, 1 00
* Hoff's Elementary Naval Tactics........................8vo, 1 50
Ingalls's Handbook of Problems in Direct Fire............8vo, 4 00
* " Ballistic Tables..................................8vo, 1 50
Lyons's Treatise on Electromagnetic Phenomena...........8vo, 6 00
* Mahan's Permanent Fortifications. (Mercur.)..8vo, half mor., 7 50
Manual for Courts-martial......................16mo, morocco, 1 50
* Mercur's Attack of Fortified Places...................12mo, 2 00
* " Elements of the Art of War.....................8vo, 4 00
Metcalf's Cost of Manufactures—And the Administration of Workshops, Public and Private............8vo, 5 00
* " Ordnance and Gunnery......................12mo, 5 00
Murray's Infantry Drill Regulations..............18mo, paper, 10
* Phelps's Practical Marine Surveying.................8vo, 2 50
Powell's Army Officer's Examiner......................12mo, 4 00
Sharpe's Art of Subsisting Armies in War......18mo, morocco, 1 50
Walke's Lectures on Explosives..........................8vo, 4 00
* Wheeler's Siege Operations and Military Mining..........8vo, 2 00
Winthrop's Abridgment of Military Law................12mo, 2 50
Woodhull's Notes on Military Hygiene..................16mo, 1 50
Young's Simple Elements of Navigation........16mo, morocco, 1 00
Second Edition, Enlarged and Revised......16mo, mor., 2 00

ASSAYING.

Fletcher's Practical Instructions in Quantitative Assaying with the Blowpipe............................12mo, morocco, 1 50
Furman's Manual of Practical Assaying..................8vo, 3 00
Miller's Manual of Assaying..............................12mo, 1 00
O'Driscoll's Notes on the Treatment of Gold Ores..........8vo, 2 00
Ricketts and Miller's Notes on Assaying..................8vo, 3 00
Wilson's Cyanide Processes..............................12mo, 1 50
" Chlorination Process..............................12mo, 1 50

ASTRONOMY.

Craig's Azimuth...4to, 3 50
Doolittle's Treatise on Practical Astronomy................8vo, 4 00
Gore's Elements of Geodesy.................................8vo, 2 50
Hayford's Text-book of Geodetic Astronomy..............8vo, 3 00
Merriman's Elements of Precise Surveying and Geodesy....8vo, 2 50
* Michie and Harlow's Practical Astronomy..............8vo, 3 00
* White's Elements of Theoretical and Descriptive Astronomy. 12mo, 2 00

BOTANY.

Baldwin's Orchids of New England..................Small 8vo, 1 50
Davenport's Statistical Methods, with Special Reference to Biological Variation.......................16mo, morocco, 1 25
Thomé and Bennett's Structural and Physiological Botany. 16mo, 2 25
Westermaier's Compendium of General Botany. (Schneider.) 8vo, 2 00

CHEMISTRY.

Adriance's Laboratory Calculations and Specific Gravity Tables. 12mo, 1 25
Allen's Tables for Iron Analysis..............................8vo, 3 00
Arnold's Compendium of Chemistry. (Mandel.) (*In preparation.*)
Austen's Notes for Chemical Students..................12mo, 1 50
Bernadou's Smokeless Powder.—Nitro-cellulose, and Theory of the Cellulose Molecule..........................12mo, 2 50
Bolton's Quantitative Analysis..............................8vo, 1 50
Brush and Penfield's Manual of Determinative Mineralogy...8vo, 4 00
Classen's Quantitative Chemical Analysis by Electrolysis. (Herrick—Boltwood.) 8vo, 3 00
Cohn's Indicators and Test-papers..........................12mo, 2 00
Craft's Short Course in Qualitative Chemical Analysis. (Schaeffer.) 12mo, 2 00
Drechsel's Chemical Reactions. (Merrill.)..............12mo, 1 25
Eissler's Modern High Explosives..........................8vo, 4 00
Effront's Enzymes and their Applications. (Prescott.)....8vo, 3 00
Erdmann's Introduction to Chemical Preparations. (Dunlap.) 12mo, 1 25
Fletcher's Practical Instructions in Quantitative Assaying with the Blowpipe..........................12mo, morocco, 1 50
Fowler's Sewage Works Analyses..........................12mo, 2 00
Fresenius's Manual of Qualitative Chemical Analysis. (Wells.) 8vo, 5 00
" Manual of Qualitative Chemical Analysis. Part I. Descriptive. (Wells.)..........................8vo, 3 00
" System of Instruction in Quantitative Chemical Analysis. (Allen.)..............................8vo, 6 00

Fuertes's Water and Public Health......................12mo, 1 50
Furman's Manual of Practical Assaying..................8vo, 3 00
Gill's Gas and Fuel Analysis for Engineers..............12mo, 1 25
Grotenfelt's Principles of Modern Dairy Practice. (Woll.)..12mo, 2 00
Hammarsten's Text-book of Physiological Chemistry. (Mandel.)
8vo, 4 00
Helm's Principles of Mathematical Chemistry. (Morgan.) 12mo, 1 50
Hinds's Inorganic Chemistry..........................8vo, 3 00
* " Laboratory Manual for Students................12mo, 75
Holleman's Text-book of Inorganic Chemistry. (Cooper.)...8vo, 2 50
" " " " Organic " (Walker and Mott.)
(*In preparation.*)
Hopkins's Oil-chemists' Handbook......................8vo, 3 00
Keep's Cast Iron....................................8vo, 2 50
Ladd's Manual of Quantitative Chemical Analysis........12mo, 1 00
Landauer's Spectrum Analysis. (Tingle.)................8vo, 3 00
Lassar-Cohn's Practical Urinary Analysis. (Lorenz.) (*In preparation.*)
Leach's The Inspection and Analysis of Food with Special Reference to State Control. (*In preparation.*)
Löb's Electrolysis and Electrosynthesis of Organic Compounds. (Lorenz.) 12mo, 1 00
Mandel's Handbook for Bio-chemical Laboratory......12mo, 1 50
Mason's Water-supply. (Considered Principally from a Sanitary Standpoint.) 3d Edition, Rewritten......8vo, 4 00
" Examination of water. (Chemical and Bacteriological.) 12mo, 1 25
Meyer's Determination of Radicles in Carbon Compounds. (Tingle.) 12mo, 1 00
Miller's Manual of Assaying...........................12mo, 1 00
Mixter's Elementary Text-book of Chemistry...........12mo, 1 50
Morgan's Outline of Theory of Solution and its Results..12mo, 1 00
" Elements of Physical Chemistry...............12mo, 2 00
Nichols's Water-supply. (Considered mainly from a Chemical and Sanitary Standpoint, 1883.)....................8vo, 2 50
O'Brine's Laboratory Guide in Chemical Analysis..........8vo, 2 00
O'Driscoll's Notes on the Treatment of Gold Ores..........8vo, 2 00
Ost and Kolbeck's Text-book of Chemical Technology. (Lorenz—Bozart.) (*In preparation.*)
* Penfield's Notes on Determinative Mineralogy and Record of Mineral Tests............................8vo, paper, 0 50
Pinner's Introduction to Organic Chemistry. (Austen.) 12mo, 1 50
Poole's Calorific Power of Fuels........................8vo, 3 00
* Reisig's Guide to Piece-dyeing.......................8vo, 25 00
Richards and Woodman's Air, Water, and Food from a Sanitary Standpoint 8vo, 2 00
Richards's Cost of Living as Modified by Sanitary Science 12mo, 1 00
" Cost of Food, a Study in Dietaries...........12mo, 1 00
* Richards and Williams's The Dietary Computer..........8vo, 1 50
Ricketts and Russell's Skeleton Notes upon Inorganic Chemistry. (Part I.—Non-metallic Elements.)..8vo, morocco, 75
Ricketts and Miller's Notes on Assaying.................8vo, 3 00
Rideal's Sewage and the Bacterial Purification of Sewage..8vo, 3 50
Ruddiman's Incompatibilities in Prescriptions.............8vo, 2 00
Schimpf's Text-book of Volumetric Analysis............12mo, 2 50
Spencer's Handbook for Chemists of Beet-sugar Houses. 16mo, mor., 3 00
" Handbook for Sugar Manufacturers and their Chemists ....16mo, morocco, 2 00
Stockbridge's Rocks and Soils..........................8vo, 2 50

* Tillman's Elementary Lessons in Heat	8vo,	1 50
* " Descriptive General Chemistry	8vo,	3 00
Turneaure and Russell's Public Water-supplies	8vo,	5 00
Van Deventer's Physical Chemistry for Beginners. (Boltwood.)	12mo,	1 50
Walke's Lectures on Explosives	8vo,	4 00
Wells's Laboratory Guide in Qualitative Chemical Analysis	8vo,	1 50
" Short Course in Inorganic Qualitative Chemical Analysis for Engineering Students	12mo,	1 50
Whipple's Microscopy of Drinking-water	8vo,	3 50
Wiechmann's Sugar Analysis	Small 8vo,	2 50
" Lecture-notes on Theoretical Chemistry	12mo,	3 00
Wilson's Cyanide Processes	12mo,	1 50
" Chlorination Process	12mo,	1 50
Wulling's Elementary Course in Inorganic Pharmaceutical and Medical Chemistry	12mo,	2 00

CIVIL ENGINEERING.

BRIDGES AND ROOFS. HYDRAULICS. MATERIALS OF ENGINEERING. RAILWAY ENGINEERING.

Baker's Engineers' Surveying Instruments	12mo,	3 00
Bixby's Graphical Computing Table	Paper, 19½ x 24¼ inches.	25
Davis's Elevation and Stadia Tables	8vo,	1 00
Folwell's Sewerage. (Designing and Maintenance.)	8vo,	3 00
Freitag's Architectural Engineering. 2d Ed., Rewritten	8vo,	3 50
French and Ives's Stereotomy	8vo,	2 50
Goodhue's Municipal Improvements	12mo,	1 75
Goodrich's Economic Disposal of Towns' Refuse	8vo,	3 50
Gore's Elements of Geodesy	8vo,	2 50
Hayford's Text-book of Geodetic Astronomy	8vo,	3 00
Howe's Retaining-walls for Earth	12mo,	1 25
Johnson's Theory and Practice of Surveying	Small 8vo,	4 00
" Stadia and Earth-work Tables	8vo,	1 25
Kiersted's Sewage Disposal	12mo,	1 25
Laplace's Philosophical Essay on Probabilities. (Truscott and Emory.)	12mo,	2 00
Mahan's Treatise on Civil Engineering. (1873.) (Wood.)	8vo,	5 00
* Mahan's Descriptive Geometry	8vo,	1 50
Merriman's Elements of Precise Surveying and Geodesy	8vo,	2 50
Merriman and Brooks's Handbook for Surveyors	16mo, mor.,	2 00
Merriman's Elements of Sanitary Engineering	8vo,	2 00
Nugent's Plane Surveying	8vo,	3 50
Ogden's Sewer Design	12mo,	2 00
Patton's Treatise on Civil Engineering	8vo, half leather,	7 50
Reed's Topographical Drawing and Sketching	4to,	5 00
Rideal's Sewage and the Bacterial Purification of Sewage	8vo,	3 50
Siebert and Biggin's Modern Stone-cutting and Masonry	8vo,	1 50
Smith's Manual of Topographical Drawing. (McMillan.)	8vo,	2 50
* Trautwine's Civil Engineer's Pocket-book	16mo, morocco,	5 00
Wait's Engineering and Architectural Jurisprudence	8vo,	6 00
	Sheep,	6 50
" Law of Operations Preliminary to Construction in Engineering and Architecture	8vo,	5 00
	Sheep,	5 50
Wait's Law of Contracts	8vo,	3 00
Warren's Stereotomy—Problems in Stone-cutting	8vo,	2 50
Webb's Problems in the Use and Adjustment of Engineering Instruments	16mo, morocco,	1 25

* Wheeler's Elementary Course of Civil Engineering........8vo, 4 00
Wilson's Topographic Surveying........................8vo, 3 50

BRIDGES AND ROOFS.

Boller's Practical Treatise on the Construction of Iron Highway Bridges..8vo, 2 00
* Boller's Thames River Bridge...................4to, paper, 5 00
Burr's Course on the Stresses in Bridges and Roof Trusses, Arched Ribs, and Suspension Bridges..............8vo, 3 50
Du Bois's Mechanics of Engineering. Vol. II.........Small 4to, 10 00
Foster's Treatise on Wooden Trestle Bridges.............4to, 5 00
Fowler's Coffer-dam Process for Piers....................8vo, 2 50
Greene's Roof Trusses...................................8vo, 1 25
" Bridge Trusses.......................................8vo, 2 50
" Arches in Wood, Iron, and Stone.....................8vo, 2 50
Howe's Treatise on Arches...............................8vo, 4 00
" Design of Simple Roof-trusses in Wood and Steel. 8vo, 2 00
Johnson, Bryan and Turneaure's Theory and Practice in the Designing of Modern Framed Structures......Small 4to, 10 00
Merriman and Jacoby's Text-book on Roofs and Bridges:
Part I.—Stresses in Simple Trusses......................8vo, 2 50
Part II.—Graphic Statics................................8vo, 2 50
Part III.—Bridge Design. Fourth Ed., Rewritten......8vo, 2 50
Part IV.—Higher Structures..............................8vo, 2 50
Morison's Memphis Bridge................................4to, 10 00
Waddell's De Pontibus, a Pocket Book for Bridge Engineers. 16mo, mor., 3 00
" Specifications for Steel Bridges..............12mo, 1 25
Wood's Treatise on the Theory of the Construction of Bridges and Roofs..8vo, 2 00
Wright's Designing of Draw-spans:
Part I.—Plate-girder Draws..............................8vo, 2 50
Part II.—Riveted-truss and Pin-connected Long-span Draws. 8vo, 2 50
Two parts in one volume.................................8vo, 3 50

HYDRAULICS.

Bazin's Experiments upon the Contraction of the Liquid Vein Issuing from an Orifice. (Trautwine.)..............8vo, 2 00
Bovey's Treatise on Hydraulics..........................8vo, 5 00
Church's Mechanics of Engineering......................8vo, 6 00
" Diagrams of Mean Velocity of Water in Open Channels paper, 1 50
Coffin's Graphical Solution of Hydraulic Problems..16mo, mor., 2 50
Flather's Dynamometers, and the Measurement of Power.12mo, 3 00
Folwell's Water-supply Engineering......................8vo, 4 00
Frizell's Water-power....................................8vo, 5 00
Fuertes's Water and Public Health.....................12mo, 1 50
" Water-filtration Works.........................12mo, 2 50
Ganguillet and Kutter's General Formula for the Uniform Flow of Water in Rivers and Other Channels. (Hering and Trautwine.)......................................8vo, 4 00
Hazen's Filtration of Public Water-supply...............8vo, 3 00
Hazlehurst's Towers and Tanks for Water-works........8vo, 2 50
Herschel's 115 Experiments on the Carrying Capacity of Large, Riveted, Metal Conduits..........................8vo, 2 00

Mason's Water-supply. (Considered Principally from a Sanitary Standpoint.)..............................8vo, 4 00
Merriman's Treatise on Hydraulics........................8vo, 4 00
* Michie's Elements of Analytical Mechanics..............8vo, 4 00
Schuyler's Reservoirs for Irrigation, Water-power, and Domestic Water-supply..............................Large 8vo, 5 00
Turneaure and Russell. Public Water-supplies............8vo, 5 00
Wegmann's Design and Construction of Dams............4to, 5 00
" Water-supply of the City of New York from 1658 to 1895 ....4to, 10 00
Weisbach's Hydraulics and Hydraulic Motors. (Du Bois.)..8vo, 5 00
Wilson's Manual of Irrigation Engineering..........Small 8vo, 4 00
Wolff's Windmill as a Prime Mover.......................8vo, 3 00
Wood's Turbines..8vo, 2 50
" Elements of Analytical Mechanics................8vo, 3 00

MATERIALS OF ENGINEERING.

Baker's Treatise on Masonry Construction..............8vo, 5 00
Black's United States Public Works..............Oblong 4to, 5 00
Bovey's Strength of Materials and Theory of Structures....8vo, 7 50
Burr's Elasticity and Resistance of the Materials of Engineering 8vo, 5 00
Byrne's Highway Construction..........................8vo, 5 00
" Inspection of the Materials and Workmanship Employed in Construction......................16mo, 3 00
Church's Mechanics of Engineering.....................8vo, 6 00
Du Bois's Mechanics of Engineering. Vol. I........Small 4to, 7 50
Johnson's Materials of Construction.............Large 8vo, 6 00
Keep's Cast Iron...8vo, 2 50
Lanza's Applied Mechanics................................8vo, 7 50
Martens's Handbook on Testing Materials. (Henning.) 2 v., 8vo, 7 50
Merrill's Stones for Building and Decoration..............8vo, 5 00
Merriman's Text-book on the Mechanics of Materials......8vo, 4 00
Merriman's Strength of Materials.......................12mo, 1 00
Metcalf's Steel. A Manual for Steel-users...............12mo, 2 00
Patton's Practical Treatise on Foundations...............8vo, 5 00
Rockwell's Roads and Pavements in France............12mo, 1 25
Smith's Wire: Its Use and Manufacture...........Small 4to, 3 00
" Materials of Machines.........................12mo, 1 00
Snow's Principal Species of Wood: Their Characteristic Properties. (*In preparation.*)
Spalding's Hydraulic Cement...............................12mo, 2 00
" Text-book on Roads and Pavements..........12mo, 2 00
Thurston's Materials of Engineering..............3 Parts, 8vo, 8 00
Part I.—Non-metallic Materials of Engineering and Metallurgy 8vo, 2 00
Part II.—Iron and Steel...................................8vo, 3 50
Part III.—A Treatise on Brasses, Bronzes and Other Alloys and Their Constituents...........................8vo, 2 50
Thurston's Text-book of the Materials of Construction.....8vo, 5 00
Tillson's Street Pavements and Paving Materials..........8vo, 4 00
Waddell's De Pontibus. (A Pocket-book for Bridge Engineers.) 16mo, morocco, 3 00
" Specifications for Steel Bridges...............12mo, 1 25
Wood's Treatise on the Resistance of Materials, and an Appendix on the Preservation of Timber..........8vo, 2 00
" Elements of Analytical Mechanics................8vo, 3 00

RAILWAY ENGINEERING.

Andrews's Handbook for Street Railway Engineers. 3x5 in. mor., 1 25
Berg's Buildings and Structures of American Railroads...4to, 5 00
Brooks's Handbook of Street Railroad Location..16mo, morocco, 1 50
Butts's Civil Engineer's Field-book............16mo, morocco, 2 50
Crandall's Transition Curve..................16mo, morocco, 1 50
" Railway and Other Earthwork Tables........8vo, 1 50
Dawson's Electric Railways and Tramways. Small 4to, half mor., 12 50
" "Engineering" and Electric Traction Pocket-book.
16mo, morocco, 4 00
Dredge's History of the Pennsylvania Railroad: (1879.). Paper, 5 00
* **Drinker's Tunneling, Explosive Compounds, and Rock Drills.**
4to, half morocco, 25 00
Fisher's Table of Cubic Yards....................Cardboard, 25
Godwin's Railroad Engineers' Field-book and Explorers' Guide.
16mo, morocco, 2 50
Howard's Transition Curve Field-book.........16mo, morocco, 1 50
Hudson's Tables for Calculating the Cubic Contents of Excavations and Embankments.........................8vo, 1 00
Nagle's Field Manual for Railroad Engineers....16mo, morocco, 3 00
Philbrick's Field Manual for Engineers........16mo, morocco, 3 00
Pratt and Alden's Street-railway Road-bed...............8vo, 2 00
Searles's Field Engineering....................16mo, morocco, 3 00
" Railroad Spiral.....................16mo, morocco, 1 50
Taylor's Prismoidal Formulæ and Earthwork.............8vo, 1 50
* Trautwine's Method of Calculating the Cubic Contents of Excavations and Embankments by the Aid of Diagrams 8vo, 2 00
* " The Field Practice of Laying Out Circular Curves for Railroads..................12mo, morocco, 2 50
* " Cross-section Sheet......................Paper, 25
Webb's Railroad Construction............................8vo, 4 00
Wellington's Economic Theory of the Location of Railways..
Small 8vo, 5 00

DRAWING.

Barr's Kinematics of Machinery..........................8vo, 2 50
* Bartlett's Mechanical Drawing..........................8vo, 3 00
Coolidge's Manual of Drawing..................8vo, paper, 1 00
Durley's Elementary Text-book of the Kinematics of Machines.
(In preparation.)
Hill's Text-book on Shades and Shadows, and Perspective..8vo, 2 00
Jones's Machine Design:
Part I.—Kinematics of Machinery......................8vo, 1 50
Part II.—Form, Strength and Proportions of Parts......8vo, 3 00
MacCord's Elements of Descriptive Geometry.............8vo, 3 00
" Kinematics; or, Practical Mechanism..........8vo, 5 00
" Mechanical Drawing..............................4to, 4 00
" Velocity Diagrams...............................8vo, 1 50
* Mahan's Descriptive Geometry and Stone-cutting.......8vo, 1 50
Mahan's Industrial Drawing. (Thompson.)...............8vo, 3 50
Reed's Topographical Drawing and Sketching.............4to, 5 00
Reid's Course in Mechanical Drawing....................8vo, 2 00
" Text-book of Mechanical Drawing and Elementary Machine Design...............................8vo, 3 00
Robinson's Principles of Mechanism.....................8vo, 3 00

Smith's Manual of Topographical Drawing. (McMillan.) .8vo, 2 50
Warren's Elements of Plane and Solid Free-hand Geometrical Drawing 12mo, 1 00
" Drafting Instruments and Operations..........12mo, 1 25
" Manual of Elementary Projection Drawing....12mo, 1 50
" Manual of Elementary Problems in the Linear Perspective of Form and Shadow..............12mo, 1 00
" Plane Problems in Elementary Geometry.......12mo, 1 25
" Primary Geometry..........................12mo, 75
" Elements of Descriptive Geometry, Shadows, and Perspective 8vo, 3 50
" General Problems of Shades and Shadows.......8vo, 3 00
" Elements of Machine Construction and Drawing..8vo, 7 50
" Problems, Theorems, and Examples in Descriptive Geometry 8vo, 2 50
Weisbach's Kinematics and the Power of Transmission. (Herrmann and Klein.) 8vo, 5 00
Whelpley's Practical Instruction in the Art of Letter Engraving 12mo, 2 00
Wilson's Topographic Surveying........................8vo, 3 50
Wilson's Free-hand Perspective........................8vo, 2 50
Woolf's Elementary Course in Descriptive Geometry..Large 8vo, 3 00

ELECTRICITY AND PHYSICS.

Anthony and Brackett's Text-book of Physics. (Magie.) Small 8vo, 3 00
Anthony's Lecture-notes on the Theory of Electrical Measurments 12mo, 1 00
Benjamin's History of Electricity......................8vo, 3 00
Benjamin's Voltaic Cell...............................8vo, 3 00
Classen's Qantitative Chemical Analysis by Electrolysis. Herrick and Boltwood.)..............................8vo, 3 00
Crehore and Squier's Polarizing Photo-chronograph......8vo, 3 00
Dawson's Electric Railways and Tramways..Small 4to, half mor., 12 50
Dawson's "Engineering" and Electric Traction Pocket-book. 16mo, morocco, 4 00
Flather's Dynamometers, and the Measurement of Power..12mo, 3 00
Gilbert's De Magnete. (Mottelay.)......................8vo, 2 50
Holman's Precision of Measurements.....................8vo, 2 00
" Telescopic Mirror-scale Method, Adjustments, and Tests ....Large 8vo, 75
Landauer's Spectrum Analysis. (Tingle.)................8vo, 3 00
Le Chatelier's High-temperature Measurements. (Boudouard—Burgess.) 12mo, 3 00
Löb's Electrolysis and Electrosynthesis of Organic Compounds. (Lorenz.) 12mo, 1 00
Lyons's Treatise on Electromagnetic Phenomena...........8vo, 6 00
* Michie. Elements of Wave Motion Relating to Sound and Light 8vo, 4 00
Niaudet's Elementary Treatise on Electric Batteries (Fishback.) 12mo, 2 50
* Parshall and Hobart's Electric Generators..Small 4to, half mor., 10 00
Ryan, Norris, and Hoxie's Electrical Machinery. (*In preparation.*)
Thurston's Stationary Steam-engines....................8vo, 2 50
* Tillman. Elementary Lessons in Heat..................8vo, 1 50
Tory and Pitcher. Manual of Laboratory Physics..Small 8vo, 2 00

LAW.

* Davis. Elements of Law..........................8vo, 2 50
* " Treatise on the Military Law of United States..8vo, 7 00
* Sheep, 7 50
Manual for Courts-martial..................16mo, morocco, 1 50
Wait's Engineering and Architectural Jurisprudence......8vo, 6 00
Sheep, 6 50
" Law of Operations Preliminary to Construction in Engineering and Architecture......................8vo, 5 00
Sheep, 5 50
" Law of Contracts..................................8vo, 3 00
Winthrop's Abridgment of Military Law..............12mo, 2 50

MANUFACTURES.

Beaumont's Woollen and Worsted Cloth Manufacture....12mo, 1 50
Bernadou's Smokeless Powder—Nitro-cellulose and Theory of the Cellulose Molecule..........................12mo, 2 50
Bolland's Iron Founder..........................12mo, cloth, 2 50
" "The Iron Founder" Supplement..............12mo, 2 50
" Encyclopedia of Founding and Dictionary of Foundry Terms Used in the Practice of Moulding....12mo, 3 00
Eissler's Modern High Explosives.......................8vo, 4 00
Effront's Enzymes and their Applications. (Prescott.)...8vo, 3 00
Fitzgerald's Boston Machinist..........................18mo, 1 00
Ford's Boiler Making for Boiler Makers..............18mo, 1 00
Hopkins's Oil-chemists' Handbook.......................8vo, 3 00
Keep's Cast Iron...8vo 2 50
Leach's The Inspection and Analysis of Food with Special Reference to State Control. (*In preparation.*)
Metcalf's Steel. A Manual for Steel-users..............12mo, 2 00
Metcalf's Cost of Manufactures—And the Administration of Workshops, Public and Private................8vo, 5 00
Meyer's Modern Locomotive Construction................4to, 10 00
* Reisig's Guide to Piece-dyeing..........................8vo, 25 00
Smith's Press-working of Metals..........................8vo, 3 00
" Wire: Its Use and Manufacture...........Small 4to, 3 00
Spalding's Hydraulic Cement.............................12mo, 2 00
Spencer's Handbook for Chemists of Beet-sugar Houses.
16mo, morocco, 3 00
" Handbook for Sugar Manufacturers and their Chemists...........................16mo, morocco, 2 00
Thurston's Manual of Steam-boilers, their Designs, Construction and Operation....................................8vo, 5 00
Walke's Lectures on Explosives..........................8vo, 4 00
West's American Foundry Practice......................12mo, 2 50
" Moulder's Text-book..............................12mo, 2 50
Wiechmann's Sugar Analysis.......................Small 8vo, 2 50
Wolff's Windmill as a Prime Mover......................8vo, 3 00
Woodbury's Fire Protection of Mills....................8vo, 2 50

MATHEMATICS.

Baker's Elliptic Functions...............................8vo, 1 50
* Bass's Elements of Differential Calculus..............12mo, 4 00
Briggs's Elements of Plane Analytic Geometry..........12mo, 1 00
Chapman's Elementary Course in Theory of Equations...12mo, 1 50
Compton's Manual of Logarithmic Computations........12mo, 1 50

Davis's Introduction to the Logic of Algebra..............8vo, **1 50**
***Dickson's College Algebra**......................Large 12mo, 1 50
Halsted's Elements of Geometry..........................8vo, 1 75
" Elementary Synthetic Geometry...............8vo, 1 50
* Johnson's Three-place Logarithmic Tables: Vest-pocket size, pap., 15
100 copies for **5 00**
* Mounted on heavy cardboard, 8 × 10 inches, 25
10 copies for **2 00**
" Elementary Treatise on the Integral Calculus. Small 8vo, 1 50
" **Curve Tracing in Cartesian Co-ordinates**......12mo, **1 00**
" Treatise on Ordinary and Partial Differential Equations ....Small 8vo, 3 50
" Theory of Errors and the Method of Least Squares 12mo, 1 50
* " **Theoretical Mechanics**.......................12mo, **3 00**
Laplace's Philosophical Essay on Probabilities. (Truscott and Emory.)...12mo, 2 00
* Ludlow and Bass. Elements of Trigonometry and Logarithmic and Other Tables..............................8vo, 3 00
" **Trigonometry. Tables published separately..Each, 2 00**
Merriman and Woodward. Higher Mathematics..........8vo, 5 00
Merriman's Method of Least Squares...................8vo, 2 00
Rice and Johnson's Elementary Treatise on the Differential Calculus ....Small 8vo, 3 00
" Differential and Integral Calculus. 2 vols. in one....................Small 8vo, 2 50
Wood's Elements of Co-ordinate Geometry................8vo, 2 00
" Trigometry: Analytical, Plane, and Spherical....12mo, 1 00

MECHANICAL ENGINEERING.

MATERIALS OF ENGINEERING, STEAM ENGINES AND BOILERS.

Baldwin's Steam Heating for Buildings..................12mo, **2 50**
Barr's Kinematics of Machinery.......................8vo, 2 50
* Bartlett's Mechanical Drawing.........................8vo, 3 00
Benjamin's Wrinkles and Recipes......................12mo, 2 00
Carpenter's Experimental Engineering...................8vo, 6 00
" Heating and Ventilating Buildings...........8vo, 4 00
Clerk's Gas and Oil Engine........................Small 8vo, 4 00
Coolidge's Manual of Drawing.................8vo, paper, 1 00
Cromwell's Treatise on Toothed Gearing..................12mo, 1 50
" Treatise on Belts and Pulleys...............12mo, 1 50
Durley's Elementary Text-book of the Kinematics of Machines. (*In preparation.*)
Flather's Dynamometers, and the Measurement of Power ..12mo, 3 00
" Rope Driving..............................12mo, 2 00
Gill's Gas an Fuel Analysis for Engineers..............12mo, 1 25
Hall's Car Lubrication.................................12mo, 1 00
Jones's Machine Design:
Part I.—Kinematics of Machinery.....................8vo, 1 50
Part II.—Form, Strength and Proportions of Parts......8vo, 3 00
Kent's Mechanical Engineers' Pocket-book....16mo, morocco, 5 00
Kerr's Power and Power Transmission..................8vo, 2 00

MacCord's Kinematics; or, Practical Mechanism..........8vo, 5 00
" Mechanical Drawing..........................4to, 4 00
" Velocity Diagrams............................8vo, 1 50
Mahan's Industrial Drawing. (Thompson.)..............8vo, 3 50
Poole's Calorific Power of Fuels.........................8vo, 3 00
Reid's Course in Mechanical Drawing....................8vo, 2 00
" Text-book of Mechanical Drawing and Elementary Machine Design..8vo, 3 00
Richards's Compressed Air..........................12mo, 1 50
Robinson's Principles of Mechanism.....................8vo, 3 00
Smith's Press-working of Metals.........................8vo, 3 00
Thurston's Treatise on Friction and Lost Work in Machinery and Mill Work..........................8vo, 3 00
" Animal as a Machine and Prime Motor and the Laws of Energetics......................12mo, 1 00
Warren's Elements of Machine Construction and Drawing..8vo, 7 50
Weisbach's Kinematics and the Power of Transmission. (Herrmann—Klein.) 8vo, 5 00
" Machinery of Transmission and Governors. (Herrmann—Klein.) 8vo, 5 00
" Hydraulics and Hydraulic Motors. (Du Bois.).8vo, 5 00
Wolff's Windmill as a Prime Mover.....................8vo, 3 00
Wood's Turbines...8vo, 2 50

MATERIALS OF ENGINEERING.

Bovey's Strength of Materials and Theory of Structures..8vo, 7 50
Burr's Elasticity and Resistance of the Materials of Engineering 8vo, 5 00
Church's Mechanics of Engineering......................8vo, 6 00
Johnson's Materials of Construction..............Large 8vo, 6 00
Keep's Cast Iron...8vo, 2 50
Lanza's Applied Mechanics..............................8vo, 7 50
Martens's Handbook on Testing Materials. (Henning.)....8vo, 7 50
Merriman's Text-book on the Mechanics of Materials....8vo, 4 00
" Strength of Materials.....................12mo, 1 00
Metcalf's Steel. A Manual for Steel-users..............12mo, 2 00
Smith's Wire: Its Use and Manufacture............Small 4to, 3 00
" Materials of Machines..........................12mo, 1 00
Thurston's Materials of Engineering..............3 vols., 8vo, 8 00
Part II.—Iron and Steel..................................8vo, 3 50
Part III.—A Treatise on Brasses, Bronzes and Other Alloys and their Constituents............................8vo, 2 50
Thurston's Text-book of the Materials of Construction....8vo, 5 00
Wood's Treatise on the Resistance of Materials and an Appendix on the Preservation of Timber..........8vo, 2 00
" Elements of Analytical Mechanics...............8vo, 3 00

STEAM ENGINES AND BOILERS.

Carnot's Reflections on the Motive Power of Heat. (Thurston.) 12mo, 1 50
Dawson's "Engineering" and Electric Traction Pocket-book. 16mo, morocco, 4 00
Ford's Boiler Making for Boiler Makers...............18mo, 1 00
Goss's Locomotive Sparks................................8vo, 2 00
Hemenway's Indicator Practice and Steam-engine Economy. 12mo, 2 00
Hutton's Mechanical Engineering of Power Plants........8vo, 5 00
" Heat and Heat-engines.......................8vo, 5 00

Kent's Steam-boiler Economy..........................8vo, 4 00
Kneass's Practice and Theory of the Injector..............8vo, 1 50
MacCord's Slide-valves..........................8vo, 2 00
Meyer's Modern Locomotive Construction................4to, 10 00
Peabody's Manual of the Steam-engine Indicator.........12mo, 1 50
" Tables of the Properties of Saturated Steam and Other Vapors..........................8vo, 1 00
" Thermodynamics of the Steam-engine and Other Heat-engines..........................8vo, 5 00
" Valve-gears for Steam-engines..................8vo, 2 50
Peabody and Miller. Steam-boilers......................8vo, 4 00
Pray's Twenty Years with the Indicator...........Large 8vo, 2 50
Pupin's Thermodynamics of Reversible Cycles in Gases and Saturated Vapors. (Osterberg.)..................12mo, 1 25
Reagan's Locomotive Mechanism and Engineering......12mo, 2 00
Rontgen's Principles of Thermodynamics. (Du Bois.)....8vo, 5 00
Sinclair's Locomotive Engine Running and Management..12mo, 2 00
Smart's Handbook of Engineering Laboratory Practice..12mo, 2 50
Snow's Steam-boiler Practice..........................8vo, 3 00
Spangler's Valve-gears..........................8vo, 2 50
" Notes on Thermodynamics..................12mo, 1 00
Thurston's Handy Tables..........................8vo, 1 50
" Manual of the Steam-engine..........2 vols., 8vo, 10 00
Part I.—History, Structure, and Theory................8vo, 6 00
Part II.—Design, Construction, and Operation..........8vo, 6 00
Thurston's Handbook of Engine and Boiler Trials, and the Use of the Indicator and the Prony Brake........8vo, 5 00
" Stationary Steam-engines..................8vo, 2 50
" Steam-boiler Explosions in Theory and in Practice..........................12mo, 1 50
" Manual of Steam-boilers, Their Designs, Construction, and Operation......................8vo, 5 00
Weisbach's Heat, Steam, and Steam-engines. (Du Bois.)..8vo, 5 00
Whitham's Steam-engine Design..........................8vo, 5 00
Wilson's Treatise on Steam-boilers. (Flather.)..........16mo, 2 50
Wood's Thermodynamics, Heat Motors, and Refrigerating Machines..........................8vo, 4 00

MECHANICS AND MACHINERY.

Barr's Kinematics of Machinery..........................8vo, 2 50
Bovey's Strength of Materials and Theory of Structures..8vo, 7 50
Chordal.—Extracts from Letters..........................12mo, 2 00
Church's Mechanics of Engineering......................8vo, 6 00
" Notes and Examples in Mechanics............8vo, 2 00
Compton's First Lessons in Metal-working..............12mo, 1 50
Compton and De Groodt. The Speed Lathe..............12mo, 1 50
Cromwell's Treatise on Toothed Gearing................12mo, 1 50
" Treatise on Belts and Pulleys................12mo, 1 50
Dana's Text-book of Elementary Mechanics for the Use of Colleges and Schools..........................12mo, 1 50
Dingey's Machinery Pattern Making...................12mo, 2 00
Dredge's Record of the Transportation Exhibits Building of the World's Columbian Exposition of 1893.....4to, half mor., 5 00
Du Bois's Elementary Principles of Mechanics:
Vol. I.—Kinematics..........................8vo, 3 50
Vol. II.—Statics..........................8vo, 4 00
Vol. III.—Kinetics..........................8vo, 3 50
Du Bois's Mechanics of Engineering. Vol. I........Small 4to, 7 50
" " " " " Vol.II........Small 4to, 10 00

Durley's Elementary Text-book of the Kinematics of Machines. (*In preparation.*)
Fitzgerald's Boston Machinist..........................16mo, 1 00
Flather's Dynamometers, and the Measurement of Power.12mo, 3 00
" Rope Driving..........................12mo, 2 00
Goss's Locomotive Sparks..........................8vo, 2 00
Hall's Car Lubrication..........................12mo, 1 00
Holly's Art of Saw Filing..........................18mo, 75
* Johnson's Theoretical Mechanics..........................12mo, 3 00
Johnson's Short Course in Statics by Graphic and Algebraic Methods. (*In preparation.*)
Jones's Machine Design:
Part I.—Kinematics of Machinery..........................8vo, 1 50
Part II.—Form, Strength and Proportions of Parts....8vo, 3 00
Kerr's Power and Power Transmission..........................8vo, 2 00
Lanza's Applied Mechanics..........................8vo, 7 50
MacCord's Kinematics; or, Practical Mechanism..........8vo, 5 00
" Velocity Diagrams..........................8vo, 1 50
Merriman's Text-book on the Mechanics of Materials......8vo, 4 00
* Michie's Elements of Analytical Mechanics..............8vo, 4 00
Reagan's Locomotive Mechanism and Engineering.......12mo, 2 00
Reid's Course in Mechanical Drawing..........................8vo, 2 00
" Text-book of Mechanical Drawing and Elementary Machine Design..........................8vo, 3 00
Richards's Compressed Air..........................12mo, 1 50
Robinson's Principles of Mechanism..........................8vo, 3 00
Ryan, Norris, and Hoxie's Electrical Machinery. (*In preparation.*)
Sinclair's Locomotive-engine Running and Management..12mo, 2 00
Smith's Press-working of Metals..........................8vo, 3 00
" Materials of Machines..........................12mo, 1 00
Thurston's Treatise on Friction and Lost Work in Machinery and Mill Work..........................8vo, 3 00
" Animal as a Machine and Prime Motor, and the Laws of Energetics..........................12mo, 1 00
Warren's Elements of Machine Construction and Drawing..8vo, 7 50
Weisbach's Kinematics and the Power of Transmission. (Herrman—Klein.) 8vo, 5 00
" Machinery of Transmission and Governors. (Herr-(man—Klein.) 8vo, 5 00
Wood's Elements of Analytical Mechanics..............8vo, 3 00
" Principles of Elementary Mechanics............12mo, 1 25
" Turbines 8vo, 2 50
The World's Columbian Exposition of 1893................4to, 1 00

METALLURGY.

Egleston's Metallurgy of Silver, Gold, and Mercury:
Vol. I.—Silver..........................8vo, 7 50
Vol. II.—Gold and Mercury..........................8vo, 7 50
** Iles's Lead-smelting..........................12mo, 2 50
Keep's Cast Iron..........................8vo, 2 50
Kunhardt's Practice of Ore Dressing in Europe..........8vo, 1 50
Le Chatelier's High-temperature Measurements. (Boudouard—Burgess.) 12mo, 3 00
Metcalf's Steel. A Manual for Steel-users..............12mo, 2 00
Smith's Materials of Machines..........................12mo, 1 00
Thurston's Materials of Engineering. In Three Parts......8vo, 8 00
Part II.—Iron and Steel..........................8vo, 3 50
Part III.—A Treatise on Brasses, Bronzes and Other Alloys and Their Constituents..........................8vo, 2 50

MINERALOGY.

Barringer's Description of Minerals of Commercial Value. Oblong, morocco, 2 50
Boyd's Resources of Southwest Virginia................8vo, 3 00
" Map of Southwest Virginia........Pocket-book form, 2 00
Brush's Manual of Determinative Mineralogy. (Penfield.).8vo, 4 00
Chester's Catalogue of Minerals...................8vo, paper, 1 00
Cloth, 1 25
" Dictionary of the Names of Minerals............8vo, 3 50
Dana's System of Mineralogy.... ......Large 8vo, half leather, 12 50
" First Appendix to Dana's New "System of Mineralogy." Large 8vo, 1 00
" Text-book of Mineralogy.........................8vo, 4 00
" Minerals and How to Study Them...............12mo, 1 50
" Catalogue of American Localities of Minerals.Large 8vo, 1 00
" Manual of Mineralogy and Petrography..........12mo, 2 00
Egleston's Catalogue of Minerals and Synonyms..........8vo, 2 50
Hussak's The Determination of Rock-forming Minerals. (Smith.) ....Small 8vo, 2 00
* Penfield's Notes on Determinative Mineralogy and Record of Mineral Tests.............................8vo, paper, 50
Rosenbusch's Microscopical Physiography of the Rock-making Minerals. (Idding's.)..............................8vo, 5 00
* Tillman's Text-book of Important Minerals and Rocks..8vo, 2 00
Williams's Manual of Lithology............................8vo, 3 00

MINING.

Beard's Ventilation of Mines...........................12mo, 2 50
Boyd's Resources of Southwest Virginia................8vo, 3 00
" Map of Southwest Virginia........Pocket-book form, 2 00
* Drinker's Tunneling, Explosive Compounds, and Rock Drills...............................4to, half morocco, 25 00
Eissler's Modern High Explosives........................8vo, 4 00
Fowler's Sewage Works Analyses........................12mo, 2 00
Goodyear's Coal-mines of the Western Coast of the United States 12mo, 2 50
Ihlseng's Manual of Mining...............................8vo, 4 00
** Iles's Lead-smelting...................................12mo, 2 50
Kunhardt's Practice of Ore Dressing in Europe............8vo, 1 50
O'Driscoll's Notes on the Treatment of Gold Ores..........8vo, 2 00
Sawyer's Accidents in Mines...............................8vo, 7 00
Walke's Lectures on Explosives...........................8vo, 4 00
Wilson's Cyanide Processes..............................12mo, 1 50
Wilson's Chlorination Process...........................12mo, 1 50
Wilson's Hydraulic and Placer Mining...................12mo, 2 00
Wilson's Treatise on Practical and Theoretical Mine Ventilation 12mo, 1 25

SANITARY SCIENCE.

Folwell's Sewerage. (Designing, Construction and Maintenance.) 8vo, 3 00
" Water-supply Engineering.......................8vo, 4 00
Fuertes's Water and Public Health.......... ....12mo, 1 50
" Water-filtration Works.......................12mo, 2 50

Gerhard's Guide to Sanitary House-inspection..........16mo, 1 00
Goodrich's Economical Disposal of Towns' Refuse...Demy 8vo, 3 50
Hazen's Filtration of Public Water-supplies...............8vo, 3 00
Kiersted's Sewage Disposal..............................12mo, 1 25
Leach's The Inspection and Analysis of Food with Special Reference to State Control. (*In preparation.*)
Mason's Water-supply. (Considered Principally from a Sanitary Standpoint. 3d Edition, Rewritten.....8vo, 4 00
" Examination of Water. (Chemical and Bacteriological.) 12mo, 1 25
Merriman's Elements of Sanitary Engineering............8vo, 2 00
Nichols's Water-supply. (Considered Mainly from a Chemical and Sanitary Standpoint.) (1883.) 8vo, 2 50
Ogden's Sewer Design...................................12mo, 2 00
* Price's Handbook on Sanitation.......................12mo, 1 50
Richards's Cost of Food. A Study in Dietaries...........12mo, 1 00
Richards and Woodman's Air, Water, and Food from a Sanitary Standpoint...................................8vo, 2 00
Richards's Cost of Living as Modified by Sanitary Science.12mo, 1 00
* Richards and Williams's The Dietary Computer..........8vo, 1 50
Rideal's Sewage and Bacterial Purification of Sewage......8vo, 3 50
Turneaure and Russell's Public Water-supplies............8vo, 5 00
Whipple's Microscopy of Drinking-water..................8vo, 3 50
Woodhull's Notes on Military Hygiene...................16mo, 1 50

MISCELLANEOUS.

Barker's Deep-sea Soundings............................8vo, 2 00
Emmons's Geological Guide-book of the Rocky Mountain Excursion of the International Congress of Geologists. Large 8vo, 1 50
Ferrel's Popular Treatise on the Winds..................8vo, 4 00
Haines's American Railway Management...............12mo, 2 50
Mott's Composition, Digestibility, and Nutritive Value of Food. Mounted chart, 1 25
" Fallacy of the Present Theory of Sound..........16mo, 1 00
Ricketts's History of Rensselaer Polytechnic Institute, 1824–1894....................................Small 8vo, 3 00
Rotherham's Emphasised New Testament...........Large 8vo, 2 00
" Critical Emphasised New Testament.......12mo, 1 50
Steel's Treatise on the Diseases of the Dog...............8vo, 3 50
Totten's Important Question in Metrology...............8vo, 2 50
The World's Columbian Exposition of 1893..............4to, 1 00
Worcester and Atkinson. Small Hospitals, Establishment and Maintenance, and Suggestions for Hospital Architecture, with Plans for a Small Hospital..................12mo, 1 25

HEBREW AND CHALDEE TEXT-BOOKS.

Green's Grammar of the Hebrew Language..............8vo, 3 00
" Elementary Hebrew Grammar.................12mo, 1 25
" Hebrew Chrestomathy.........................8vo, 2 00
Gesenius's Hebrew and Chaldee Lexicon to the Old Testament Scriptures. (Tregelles.)........Small 4to, half morocco, 5 00
Letteris's Hebrew Bible..................................8vo, 2 25

Printed in Dunstable, United Kingdom

91220010R00127